Instructor's Manual with Prepared Tests for

TECHNICAL MATHEMATICS
FOURTH EDITION

JACQUELINE C. AUSTIN
Miami-Dade Community College

JACK G. GILL
Miami-Dade Community College

 SAUNDERS COLLEGE PUBLISHING

Philadelphia   New York   Chicago
San Francisco   Montreal   Toronto
London   Sydney   Tokyo

Printed in the United States of America

Instructor's Manual with Prepared Tests for TECHNICAL MATHEMATICS, 4th. ed.

ISBN 0-03-0132320

Introduction

Purpose

- To review the basic arithmetic processes with whole numbers, fractions, decimals, and percents.

- To introduce exponents.

- To introduce scientific notation.

- To introduce the metric system.

- To emphasize the importance of the calculator in performing numerical calculations.

Chapter 1 includes seventeen sections. They are: 1.1 Addition and Subtraction of Whole Numbers; 1.2 Multiplication and Division of Whole Numbers; 1.3 Exponents; 1.4 Prime Numbers and Prime Factorization; 1.5 Fractions; 1.6 Equivalent Fractions; 1.7 Fractions in Lowest Terms; 1.8 Lowest Common Denominator; 1.9 Addition and Subtraction of Fractions; 1.10 Multiplication and Division of Fractions; 1.11 Decimals and Rounding; 1.12 Addition and Subtraction of Decimals; 1.13 Multiplication and Division of Decimals; 1.14 Hand-Held Calculators; 1.15 Fraction and Decimal Equivalences; 1.16 Percent; 1.17 Measures. These seventeen sections are followed by 119 review problems and a Self-Evaluation test.

What Makes This Book Special

- A slow and gradual development of even the most basic arithmetic processes for those students in need.

- The inclusion of numerous technology applications of the topics presented.

Chapter Objectives

(1.1)	(1) Add whole numbers.
	(2) Subtract whole numbers.
(1.2)	(3) Multiply whole numbers.
	(4) Divide whole numbers.
(1.3)	(5) Express numbers in exponential form as integers.
(1.4)	(6) Find the prime factorization of whole numbers.
(1.5)	(7) Express improper fractions as mixed numbers.
	(8) Express mixed numbers as improper fractions.
(1.6)	(9) Express fractions as equivalent fractions with a given denominator.
(1.7)	(10) Reduce fractions to lowest terms.
(1.8)	(11) Find the lowest common denominator of two or more fractions.
(1.9)	(12) Add fractions.
	(13) Subtract fractions.
(1.10)	(14) Multiply fractions.
	(15) Divide fractions.
(1.11)	(16) Round decimals.
	(17) Identify significant digits.
(1.12)	(18) Add decimals.
	(19) Subtract decimals.
(1.13)	(20) Multiply decimals.
	(21) Divide decimals.
	(22) Express numbers in scientific notation.
(1.14)	(23) Use a calculator to perform numerical calculations.

(1.15)   (24) Express decimals as fractions.
             (25) Express fractions as decimals,
(1.16)   (26) Express a decimal or a fraction as a percent.
             (27) Express a percent as a decimal or a fraction.
(1.17)   (28) Convert a quantity in one unit of measure to an equivalent quantity in another unit of measure.

Lecture Notes and Suggestions

- Cover the early sections in the chapter quickly but refer deficient students to the appropriate sections for additional reinforcement.

- Encourage students to solve problems using the calculator wherever possible.

Suggested Assignments

- Use the even-numbered exercises to reinforce the lecture.

- Assign all even-numbered exercises.

- Assign all word problems.

- Use the Chapter Review for students who need more help before a Chapter Test.

Section 1.1

2. 905   4. 1175   6. 6895   8. 2645   10. 597 kg   12. 126 min or 2 hr 6 min

14. 10,813 kw/hr   16. 3489 km   18. 1750   20. 95   22. 18 + 15 = 33; 27 + 6 = 33

24. 6874 + 2659 = 9533; 2659 + 6874 = 9533   26. 273   28. 415   30. 2001

32. 2191   34. 286 km   36. 363,252   38. 9 ft   40. 1438 km   42. 1509

44. 129   46. 279   48. 132   50. 93

Section 1.2

2. 1564   4. 21,049   6. 21,120   8. 627,300   10. 3,738,220   12. 2400 km

14. 140 hr 40 min 15 sec   16. 1044 km^2 18. 87 sq m.   20. 2364   22. 852 kg

24. $9276   26. 30 x (45 + 20) = (30 x 45) + (30 x 20)
30 x 65 = 1350 + 600
1950 = 1950

28. (19 + 31) x 15 = (19 x 15) + (31 x 15)   30. (23 x 19) + (23 x 41) = 23 x (19 + 41)
50 x 15 = 285 + 465   437 + 943 = 23 x 60
750 = 750   1380 = 1380

32. (510 x 25) + (490 x 25) = (510 + 490) x 25
12,750 + 12,250 = 1000 x 25
25,000 = 25,000

34. 9 R 3   36. 43 R 4   38. 12 R 31   40. 183   42. 514 kg   44. 239 km

46. 84   48. 45 km/hr   50. 7 km/L   52. $81   54. 86   56. 168   58. 879

60. 2598   62. 122

Section 1.3

2. 2^3 4. 9^1 6. 4^3 8. 1   10. 81   12. 3   14. 1   16. 80   18. 3456

20. $(5 \times 10^2) + (1 \times 10^1) + (6 \times 10^0)$ 22. $(6 \times 10^3) + (9 \times 10^2) + (5 \times 10^1) + (4 \times 10^0)$

24. $(2 \times 10^5) + (5 \times 10^4) + (4 \times 10^3) + (7 \times 10^2) + (8 \times 10^1) + (6 \times 10^0)$ 26. 400

28. 0.49

Section 1.4

2. $2^2 \times 3$ 4. $3^2 \times 5$ 6. 5×11^2 8. $2^4 \times 3 \times 5^2$ 10. $5^2 \times 29$

12. $2^3 \times 5^2 \times 7$ 14. $2 \times 5^2 \times 13$

Section 1.5

2. $1\frac{2}{3}$ 4. $2\frac{4}{5}$ 6. $2\frac{1}{2}$ 8. $12\frac{1}{5}$ 10. $3\frac{3}{55}$ 12. $\frac{15}{4}$ 14. $\frac{37}{7}$

16. $\frac{148}{11}$ 18. $\frac{851}{27}$ 20. $\frac{38,117}{100}$ 22. $\frac{7}{4}$ lbs   24. $35\frac{5}{8}$ in.

26. $8\frac{3}{7}$ in. long, $2\frac{1}{3}$ in. wide, and $3\frac{2}{5}$ in. thick

Section 1.6

2. $\frac{2}{10}$ 4. $\frac{10}{12}$ 6. $\frac{7}{21}$ 8. $\frac{198}{143}$ 10. $\frac{3015}{207}$ 12. $\frac{24}{64}$ 14. $\frac{12}{30}$ 16. $\frac{184}{64}$

Section 1.7

2. $\frac{1}{3}$ 4. $\frac{1}{2}$ 6. $\frac{3}{11}$ 8. $\frac{16}{9}$ 10. $\frac{44}{97}$ 12. $\frac{25}{64}$ 14. $\frac{27}{8}$

Section 1.8

2. $10, \frac{2}{10}, \frac{2}{10}$ 4. $6, \frac{4}{6}, \frac{3}{6}$ 6. $18, \frac{14}{18}, \frac{12}{18}, \frac{3}{18}$ 8. $24, \frac{9}{24}, \frac{18}{24}, \frac{20}{24}$

10. $126, \frac{63}{126}, \frac{70}{126}, \frac{90}{126}$ 12. $84, \frac{18}{84}, \frac{7}{84}$ 14. $\frac{3}{72}$ and $\frac{4}{72}$

Section 1.9

2. $\frac{1}{2}$ 4. $1\frac{1}{4}$ 6. $1\frac{5}{9}$ 8. $1\frac{1}{4}$ 10. $\frac{5}{12}$ 12. $\frac{2}{5}$ 14. $1\frac{1}{6}$ 16. $1\frac{11}{18}$

18. $\left(\frac{3}{7} + \frac{5}{9}\right) + \frac{10}{14} = 1\frac{44}{63}; \quad \frac{3}{7} + \left(\frac{5}{9} + \frac{10}{14}\right) = 1\frac{44}{63}$ 20. $\frac{4}{45}$ 22. $\frac{1}{6}$ 24. $6\frac{3}{4}$

26. $1\frac{4}{5}$ 28. $5\frac{1}{30}$ 30. $9\frac{19}{48}$ 32. $\frac{15}{16}$ in.   34. $37\frac{3}{14}\ell$ 36. $8\frac{9}{32}$ cm

38. $\frac{15}{16}$ in.   40. $9\frac{3}{8}$ in.

Section 1.10

2. $\frac{4}{15}$ 4. $\frac{5}{18}$ 6. $\frac{1}{10}$ 8. 27 10. $\frac{15}{2}$ 12. 1 14. $\frac{14}{9}$ 16. $\frac{36}{25}$

18. $\frac{7}{32}$ 20. $\frac{5}{12}$ 22. 4 24. $\frac{1}{4}$ 26. 7 28. $\frac{3}{16}$ 30. 40 32. 60

34. $80\frac{11}{25}$ 36. $333\frac{1}{3}$ 38. $89\frac{5}{19}$

Section 1.11

2. six and eight hundred sixteen thousandths   4. six hundred twenty-nine and four thousandths   6. nine and four hundred eight thousandths   8. three hundred and one thousandths   10. sixty-eight and twenty hundredths   12. 403.02   14. 0.65

16. 8024.623   18. 69.05   20. 54.501   22. 1   24. 3   26. 3   28. 1

30. 4   32. 3   34. 600   36. 50   38. 27.80   40. 4   42. 110   44. 5000 ft

46. 0.275 in.

Section 1.12

2. 1.6   4. 20.89   6. 264   8. 0.4   10. 3   12. 17   14. 14   16. 4.6

18. 12.0   20. 10.77   22. 703.8   24. 9.1°   26. $12.84   28. 422.5

Section 1.13

2. 9   4. 0.54   6. 0.79   8. 0.018   10. 500   12. 0.03   14. 0.2

16. 0.06   18. 0.6   20. 21   22. 650   24. 0.6   26. $.17   28. $2.35

30. 33.5   32. $15.00   34. $5.20   36. $13.53   38. 6×10^4 40. 7.82×10^3

42. 1.8×10^3 44. 6.7   46. 4.1   48. 28.7

Section 1.14

2. 5542   4. 2568   6. 17,283   8. 3221.61   10. 1680   12. 242,970,624

14. 23,680   16. 16   18. 12,9615   20. 40.59556   22. 7798   24. 1.28

Section 1.15

2. $\frac{4}{5}$ 4. $\frac{19}{20}$ 6. $\frac{1}{5}$ 8. $\frac{301}{100}$ 10. $\frac{41,379}{1000}$ 12. $\frac{29}{500}$ 14. $\frac{8401}{20}$ 16. 0.25

18. 0.625   20. 0.6   22. $\frac{1}{20}$ 24. 1.8

Section 1.16

2. 47%   4. 73.2%   6. 43.67%   8. 0.05%   10. 400%   12. 60%   14. 33.3%

16. 16.67%   18. 666.67%   20. 0.57   22. 0.007   24. 0.437   26. 0.058

28. 0.06375   30. 4.45   32. $\frac{1}{4}$ 34. $\frac{259}{400}$ 36. $\frac{37}{500}$ 38. 1   40. $3\frac{1}{2}$

42. 126%, $\frac{63}{50}$ 44. 5%   46. 0.92   48. $\frac{7}{50}$

2. 3.7 ℓ 4. 0.000883 km   6. 281,600 mg   8. 450 cm   10. 0.44 g

12. 33,710 mm^2 14. 0.0515 m^2 16. 17,170 m^3 18. 162 in.   20. 2.9 ℓ

22. 79 cm   24. 88.2 mi   26. 1.33 ft   28. 9.29 ft   30. 0.750 ft   32. $14\frac{5}{9}$ ft^2

34. 763 ft   36. 0.0247 m   38. 3.175 cm

CHAPTER 2

APPLIED GEOMETRY

Introduction

Purpose

- To introduce the student to the important vocabulary of geometry.

- To develop the necessary geometry skills to cope with the variety of technical situations that the student might encounter.

- To help the student work confidently with problems involving spatial relations.

Chapter 2 includes six sections. They are: 2.1 Points and Lines; 2.2 Angles; 2.3 Triangles; 2.4 Polygons; 2.5 Circles; 2.6 Solids. The six sections are followed by 31 review problems and a Self-Evaluation Test.

What Makes This Book Special

- The development of informal geometry with the assumption that the student has very little background in this area.

Chapter Objectives

(2.1) (1) Distinguish between lines, line segments, and rays.
       (2) Determine whether lines are parallel, intersecting, or coincident.
(2.2) (3) Recognize different kinds of angles.
(2.3) (4) Find the perimeter of a triangle.
       (5) Find the area of a triangle.
(2.4) (6) Find the area of a given polygon.
       (7) Find the perimeter of a given polygon.
(2.5) (8) Find the circumference of a circle.
       (9) Find the area of a circle.
(2.6) (10) Find the surface area of a given solid.
       (11) Find the volume of a given solid.

Lecture Notes and Suggestions

- Have students master all vocabulary terms.

- Have students use their calculators to solve problems.

Suggested Assignments

- Use even-numbered exercises to reinforce the lecture.

- Assign even-numbered exercises.

- Have students work through the Chapter Review before they encounter the chapter test.

CHAPTER 2 - ANSWERS TO EVEN-NUMBERED PROBLEMS

Section 2.1

2. 1   **4.** $\overrightarrow{AX}$, $\overrightarrow{AY}$, $\overrightarrow{XY}$ or $\overrightarrow{XA}$, $\overrightarrow{YA}$, $\overrightarrow{YX}$ **6.** $\overrightarrow{XU}$ **8.** line n   **10.** all points

12. one   **14.** $\overline{AB}$, $\overline{AC}$, $\overline{BC}$

Section 2.2

2. 67°   **4.** 28°23'   **6.** 81°16' 35"   **8.** acute   **10.** obtuse   **12.** obtuse

14. obtuse   **16.** 91°   **18.** 129°11'44"   **20.** 18°32' 7"   **22.** 1°   **24.** $\angle$n = 118°

26. $\angle$Z = 25°, $\angle$X = 119°, $\angle$Y = 155°   **28.** $\angle\mu$ = 57°, $\angle\sigma$ = 123°   **30.** 65°   **32.** 180°

34. 75°   **36.** 140°   **38.** $\angle$UTV and $\angle$QSR, $\angle$PTU and $\angle$RSW

Section 2.3

2. $\angle$BAD or $\angle$DCA   **4.** $\triangle$ CDE   **6.** $\triangle$ AFB, $\triangle$ AFD   **8.** 10.3 m, 3.4 m^2 **10.** 6 ft 9 in.

12. 3,515 ft^2 **14.** 0.25 ft^2

Section 2.4

2. rectangle, 14 cm, 11 cm^2 **4.** parallelogram, 18 m, 16 m^2 **6.** hexagon, 6 m, 2.6 m^2

8. hexagon, 24 cm, 36 cm   **10.** polygon, 31 m, 22.25 m^2 **12.** hexagon, 4 m, 0.75 m^2

14. decagon, 4 cm, 0.63 cm^2 **16.** 98 ft   **18.** 58 ft^2 **20.** 60.4 ft

Section 2.5

2. 31.4 m, 78.5 m^2 **4.** 27.3 cm, 59.1 cm^2 **6.** 250.7 m^2 **8.** 84 m^2 **10.** 14.8 cm^2

12. 3.44 in^2 **14.** 25.12 m   **16.** 244.9 cm   **18.** 49,062.5 in^2

Section 2.6

2. 504 m^2, 720 m^3 **4.** 600 cm^2, 1000 cm^3 **6.** 1640 m^2, 3790 m^3 **8.** 940 m^2, 1080 m^3

10. 845 m^2, 1670 m^3 **12.** 38,000 m^3, 700,000 m^3 **14.** 65 m^3 **16.** 11,530 in.

CHAPTER 3

BASIC ALGEBRAIC OPERATIONS

Introduction

Purpose

- To develop the basic algebraic skills of adding, subtracting, multiplying, and dividing real numbers.

- To use the basic algebraic skills to add, subtract, multiply, and divide polynomials.

- To use the number line for comprehension of the rules for addition and multiplication of real numbers.

Chapter 3 includes nine sections. They are: 3.1 Real Numbers; 3.2 Addition of Real Numbers; 3.3 Subtraction of Real Numbers; 3.4 Addition and Subtraction of Polynomials; 3.5 Multiplication and Division of Real Numbers; 3.6 Multiplication of Polynomials; 3.7 Division of Polynomials; 3.8 Symbols of Grouping; 3.9 Evaluation of Algebraic Expressions.
The nine sections are followed by 90 review problems and a Self-Evaluation Test.

What Makes This Book Special

- The development of the Real Numbers.

- The repetition and continuation of subjects introduced in Chapter 1 such as ordering of numbers, properties, order of operations, use of the calculator, and symbols of grouping. These topics were defined with respect to the set of whole numbers, and are defined in Chapter 3 with respect to the set of real numbers.

- Addition and subtraction of real numbers followed by the addition and subtraction of polynomials.

Chapter Objectives

(3.1)	(1) Determine the absolute value of a real number.
	(2) Determine the negative of a real number.
(3.2)	(3) Find the sum of real numbers.
(3.3)	(4) Subtract real numbers.
(3.4)	(5) Add and subtract polynomials.
(3.5)	(6) Multiply and divide real numbers.
(3.6)	(7) Multiply two monomials.
	(8) Multiply a polynomial by a monomial.
(3.7)	(9) Divide a monomial by a monomial.
	(10) Divide a polynomial by a monomial.
(3.8)	(11) Remove symbols of grouping.
(3.9)	(12) Evaluate algebraic expressions and formulas for given values of variables.

Lecture Notes and Suggestions

- Have students add using the number line whenever they have difficulty with the rules for addition and multiplication of real numbers.

- Use Examples 1, 2, and 3 of Section 3.3 for drill on subtraction.

- Use Chapter 1 if any students need reinforcement on the order of operations, the

9

calculator, or the commutative, associative and distributive properties.

Suggested Assignments

- Use even-numbered exercises in class to reinforce the lecture.

- Assign even-numbered exercises.

- Use the Chapter Review for students who need more help before a Chapter Test.

CHAPTER 3 – ANSWERS TO EVEN–NUMBERED PROBLEMS

Section 3.1

2. +4.5 ft   **4.** –$275.68   **6.** –4 $\frac{1}{2}$ cu ft/min   **8.** +3   **10.** 0   **12.** 6

14. –3,002   **16.** <   **18.** >   **20.** >   **22.** –16   **24.** $\frac{7}{8}$ **26.** 50   **28.** –144.99

30. 3   **32.** 0   **34.** 1.5   **36.** – $\frac{7}{10}$, – $\frac{2}{5}$, $\frac{3}{10}$, $\frac{3}{5}$ **38.** – $\frac{2}{5}$, – $\frac{1}{3}$, – $\frac{3}{15}$, – $\frac{1}{30}$

40. –5.9, –5.16, 5.6, 5.76   **42.** –2 $\frac{4}{5}$, –2.7, –2.12, –2 $\frac{7}{25}$

Section 3.2

2.
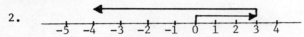

4.

6.

8.
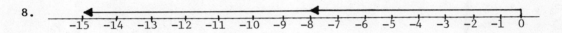

10. –13   **12.** 11   **14.** 17   **16.** –26   **18.** 5   **20.** –4   **22.** 13   **24.** –4999

26. –4.75   **28.** 8   **30.** –14   **32.** 2   **34.** $\frac{1}{10}$ **36.** $\frac{1}{14}$ **38.** 20 kg

40. profit $13,390   **42a.** Yes   **42b.** Yes   **42c.** Yes   **44.** +7.6 pounds

46. 211.14   **48.** 50.686   **50.** –0.3551

Section 3.3

2. –5   **4.** –11   **6.** 1   **8.** 4   **10.** 0   **12.** –11   **14.** 0   **16.** 0   **18.** 10

20. –19   **22.** –13,218   **24.** –2.85   **26.** 80%   **28.** $\frac{-7}{30}$ **30.** $\frac{-1}{42}$ **32.** 71.35

34. –11.981   **36.** 8   **38.** 10   **40.** 14   **42.** –25   **44.** –15   **46.** $449.84

48. –80.5 kg   **50.** 33A   **52.** 55 pounds   **54.** 71.35   **56.** –11.981   **58.** –4239

Section 3.4

2. $9Y$ **4.** $-2b$ **6.** $-6R$ **8.** 0 **10.** $19y$ **12.** $-A$ **14.** $5X - 1$ **16.** $N - 7$

18. $-22X - Y$ **20.** $-7B - 1$ **22.** $-20V + 19$ **24.** $2r + 2s - 5t$ **26.** $9X^2 - 4.7X^3$

28. $-5Y - 5Y^2 + 2Y^3$ **30.** $2a + 2c$ **32.** $4 + a - a^2 + 2a^3$

34. $\frac{17}{10}R^3 + \frac{27}{10}R^2 + \frac{11}{9}R + \frac{1}{12}$ **36.** $-2P_1 - 8P_0 + 2P$ **38.** $xy^2 + 5xy - 4x^2y$

40. $21Z_0 - 8Z_1 - 18$ **42.** $6R^2 + 9R + 21$ **44.** $20M_1 - 16M_0 - 2M$ **46.** $-4g + 1$

48. $S^2 + 7S + 1$ **50.** $4W + 14$ **52.** $210 - 2a$ **54.** $20R$ **56.** $30a$ **58.** $10X + 20$

60. $20X + 29$ **62.** $10.95f - 6.642$ **64.** $0.356a - 6.15b$

Section 3.5

2. -18 **4.** -21 **6.** 6 **8.** -81 **10.** 0 **12.** -672 **14.** 121 **16.** -1600

18. 6 **20.** 0 **22.** $\frac{1}{2}$ **24.** $\frac{3}{8}$ **26.** -2 **28.** -4 **30.** 0 **32.** 57

34. $-\frac{25}{9}$ **36.** $-\frac{5}{12}$ **38.** -90 **40.** -78 **42.** 18 **44.** -20 **46.** 20

48. -3 **50.** 0 **52.** 25 **54.** $-7°$ **56.** 5 joules **58.** 144 ft-1b **60.** $-3\frac{1}{2}$

62. 94 **64.** -154.13 **66.** 200.61

Section 3.6

2. a^7 **4.** R^7 **6.** c^{10} **8.** 5^5 **10.** $-15T^3$ **12.** $8M^2$ **14.** $-12q^4$

16. $-48a^2b^4c^2$ **18.** $0.0048X^2Y^2Z$ **20.** $\frac{3}{7}D^3$ **22.** $-24D^8$ **24.** $-300A^5B^6$

26. $15D^8T^{10}$ **28.** $-10Y - 12$ **30.** $8N^2 - N$ **32.** $2y^2 + 4y$ **34.** $6b^3c^2 - 3b^2c^3d$

36. $20X^3 - 35X^2$ **38.** $0.01R_1^2 - 0.06RR_1$ **40.** $-30Y^2 + 42Y - 48$

42. $\frac{7}{9}U^4 + \frac{7}{8}U^2 - 21U$ **44.** $9S_0^2 - 3S_0T_1$ **46.** $0.00012\,WV^2 - 0.0066\,W^2V$

48. $-28.7\,Z^5 + 42Z^4 + 11.2\,Z^3$ **50.** $16R - 19$ **52.** $14H + 26K - 54$ **54.** $2a^2 - 6a$

56. $3\,IE - 4E$ **58.** $36\pi r^3$ **60.** $\frac{1}{6}\pi a^3$ **62.** $-27336\,Y^{16}$

64. $-62.7\,b^3c^2 + 67.65\,b^3c^3$

Section 3.7

2. a^6 **4.** $\frac{1}{R^4}$ **6.** $\frac{1}{e^7}$ **8.** Y^3 **10.** R **12.** j^4 **14.** a^5 **16.** c

18. $\frac{1}{d^4}$ **20.** $\frac{1}{r}$ **22.** $\frac{1}{j}$ **24.** $\frac{1}{d^5}$ **26.** $\frac{8}{b^3}$ **28.** $24a^3$ **30.** $\frac{8}{m^7}$ **32.** $-\frac{1}{4X}$

34. $-\frac{b^2}{4}$ **36.** $12DE^3$ **38.** $-\frac{j^2}{k^2}$ **40.** $4Z$ **42.** $-0.9a^6$ **44.** $-\frac{35r^2t^4}{s^2}$

46. $2R^2 - 4$ 48. $-7k^2 + 5k$ 50. $4d^2 + 2d + 1$ 52. $10xy + 2y^7$

54. $3p^2q - 4pq^2 - 18$ 56. $-CD - 4C^2 + \dfrac{1}{D}$ 58. $2T^3 - \dfrac{R^2}{T} + \dfrac{3}{2}$ 60. $T = \dfrac{WL}{NV}$

62. $(x + 6)$ ft/sec^2 64. $8r - 11$ 66. $-214\,Y^3$ 68. $-\dfrac{0.33\,Z^4}{V^2}$

Section 3.8

2. $11 + 5a$ 4. $13D - 14$ 6. $-10D + 2$ 8. $1 - 12R$ 10. $-9ab + 6B$

12. $6d - 48$ 14. $-18T + 8$ 16. $8M - 10$ 18. $13 - 8p$ 20. $2r - 30s + 49$

22. $2a - 4$ 24. $-129p - 105$ 26. $-16X - 12$ 28. $75{,}000$ sq ft $E_o - 75{,}000$ sq ft E_f

30. $A = 3\pi\,a(2a + 3)^2$ 32. $35 + 10X$ 34. $66895b^2 - 135065\,b$ 36. $-2269\,T + 6916$

Section 3.9

2. -1 4. 1 6. 3 8. 4 10. -3 12. -1 14. $-16/5$ 16. 4

18. 20 20. -4 22. 6 24. -36 26. 7 28. -45 30a. 20 30b. 90

32a. 389 cm   32b. 236.48 cm   34a. 58.125 kg/cm^2 34b. 21.28 kg/cm^2

36a. 0.065 cm   36b. 0.070 cm   38a. 100 38b. 200 40. 139 sq ft

42. -5.1381 44. -34.2 46. -8.54

CHAPTER 4

LINEAR EQUATIONS

Introduction

Purpose

- To use the operations of addition, subtraction, multiplication, and division of real numbers to solve linear equations.

- To graph the solutions of linear inequalities.

- To use the procedures taught for solving linear equations in one variable to solve linear equations containing more than one variable.

- To develop the skill of translating verbal statements into algebraic statements in order to be able to solve word problems.

- To use proportions to solve problems involving direct, inverse, and joint variation.

Chapter 4 includes eight sections. They are: 4.1 Equations; 4.2 Solving Linear Equations; 4.3 Inequalities; 4.4 Solving Literal Equations; 4.5 Algebraic Statements; 4.6 Word Problems; 4.7 Ratio and Proportion; 4.8 Variation.
The eight sections are followed by 80 review problems and a Self-Evaluation Test.

What Makes This Book Special

- The development of solving a linear equation in one variable to solving a linear inequality in one variable to solving linear equations with more than one variable.

- The stress placed on translating a verbal statement to an algebraic statement before working on word problems.

- The use of proportions as an alternate method for solving word problems.

Chapter Objectives

(4.1) (1) Solve linear equations.
(4.2) (2) Solve linear equations that contain parentheses, fractions, and decimals.
(4.3) (3) Solve inequalities and graph the solutions on the real number line.
(4.4) (4) Solve literal equations.
(4.5) (5) Translate verbal statements into algebraic expressions and equations.
(4.6) (6) Solve word problems.
(4.7) (7) Solve problems using proportions.
(4.8) (8) Solve problems using direct, inverse, or joint variation.

Lecture Notes and Suggestions

- In Exercise 4.1, have students write out the reasons for each step used in solving an equation, as done in the examples.

- Follow the practice of written explanations in Section 4.4 when solving literal equations.

- Use the list of phrases and algebraic expression in Section 4.5 as a drill.

14

Suggested Assignments

- Have students make up a list of phrases and algebraic explanations when working on Exercise 4.5.

- Have students make up word problems after working Exercise 4.6.

- Assign odd-numbered exercises.

Section 4.1

2. $N = 13$ 4. $a = 6$ 6. $D = -7$ 8. $T = -6$ 10. $Z_1 = 0$ 12. $T_0 = -2.72$

14. $b = -1$ 16. $m = 2$ 18. $x = -18$ 20. $D = .87$ 22. $K = 2$ 24. $y = 14$

26. $k = -8$ 28. $a = -6$ 30. $b = \frac{7}{2}$ 32. $R_0 = -18$ 34. $Z_1 = 30$ 36. $x = 18$

38. $T = \frac{3}{2}$ 40. $I_0 = -\frac{5}{14}$ 42. $b = \frac{3}{35}$ 44. $P = -1$ 46. $s = 2$ 48. $R = -12$

50. $x = -100$ 52. $B = \frac{40}{3}$ 54. $m = 243$ 56. $y = -6$ 58. $b = -1$ 60. $D = -18$

62. $a = -18$ 64. $V = -\frac{8}{7}$ 66. $Y = 9.603$ 68. $j = -2.285$

Section 4.2

2. $d = 5$ 4. $y = 5$ 6. no solution 8. $x = 3$ 10. $D = 0$ 12. $b = 2$

14. no solution 16. $M = 7$ 18. $R_1 = 1$ 20. $s = 2$ 22. $I = -\frac{5}{4}$

24. $y = -30$ 26. all real numbers 28. $x = 7$ 30. $k = 3$ 32. $b = 0$

34. $x = -22$ 36. $b = .4$ 38. $x = 5860$ 40. $W = -7$ 42. $T = -\frac{2}{3}$

44. $y = \frac{13}{12}$ 46. $X = 0.029$ 48. $X = 14.321$

Section 4.3

2. $x < 3$

4. $Y \leq 4$

6. $A < 10$

8. $f \geq -4$

10. $n > -1$

12. $R \leq -4$

14. $x > \frac{1}{3}$

16. $A \leq 8$

18. x $<$ 18

20. x $>$ 0

22. x $\geq \dfrac{5}{3}$

24. x $\leq$ 9

26. t $\leq$ 5

28. M $\leq$ 6

30. X $\geq$ 2

32. t $>$ 2.04 min   **34.** 4 hours   **36.** 37.2 mi   **38.** 3 $>$ 2.1   **40.** M $>$ -1.6

Section 4.4

2. $R = D/t$ **4.** $t = I/pr$ **6.** $t = \dfrac{L - L_0}{L_0 a}$ **8.** $a = \dfrac{V_2 - V_1}{t}$ **10.** $V_1 - \dfrac{P_1 V_2}{P_2}$

12. $h = \dfrac{2A}{b}$ **14.** $W = \dfrac{P - 2\ell}{2}$ **16.** $P = \dfrac{WR - H}{2}$ **18.** $D = \dfrac{2 + N}{P}$ **20.** $t = \dfrac{pD}{2s}$

22. $V_0 = 2V - V_t$ **24.** $h = \dfrac{2A}{a + b} = 8$ **26.** $t = \dfrac{V - V_0}{a} = 4$

28. $P_1 = \dfrac{P_2 T_1}{T_2} = 30$ **30a.** $m = \dfrac{2k}{v^2} = 40$ **30b.** 8   **32.** $f = ma = 7412$ kg m/sec^2

34. $V = V_0 + at = 104.1$ **36.** $A = r^2 = 42.3$ cm^2

Section 4.5

2. $-6 + X$ **4.** $X + 13$ **6.** $17 - X$ **8.** $\dfrac{X}{14}$ **10.** $-3 + \dfrac{1}{2}X$ **12.** $27X$

14. $30 + 2X$ **16.** $2(X + 2)$ **18.** $9 + 3X$ **20.** $8 + 3X$ **22.** $-6 + X = 13, X = 19$

24. $X + 13 = 39, X = 26$ **26.** $17 - X = 1, X = 16$ **28.** $\dfrac{X}{14} = -1, X = -14$

30. $-3 + \dfrac{1}{2}X = 7, X = 20$ **32.** $27X = -81, X = -3$ **34.** $30 + 2X = 14, X = -8$

36. $2(X + 2) = 2, X = -1$ **38.** $9 + 3X = -9, X = -6$ **40.** $8 + 3X = X, X = -4$

42. $X + (-4) = 2X, X = -4$ **44.** $5X + 4 = X, X = -1$ **46.** $X - 8 = -X, X = 4$

48. $L = D - (A + B)$ **50.** $L = \sqrt{r_v^2 + r_v^2}$ **52.** $601.44 - X = 204.56, X = 396.88$

54. $6.5X = 27.3, X = 4.2$

Section 4.6

2. 64 cm x 24 cm   **4.** $2\frac{2}{11}$ ft, $5\frac{9}{11}$ ft   **6.** 28 cm, 32 cm   **8.** 8 cm

10. 5 cm x 10 cm   **12.** 250 ohms   **14.** 3 oz   **16.** 7.2 ml   **18.** 400 ℓ

20. 12 ft, 48 ft, 15 ft   **22.** 34,285.7 lb   **24.** A = 31°, B = 62°, C = 87°

26. 47 and 59 teeth   **28.** 3.2V, 5.1V, 9.6V   **30.** bases are 10.5 cm, 20.5 cm; sides are 5.5 cm   **32.** 1.24m, 2.48m, 3.44m   **34.** 31.97 ℓ

Section 4.7

2. $\frac{3}{4}$ **4.** $\frac{1}{2}$ **6.** $\frac{13}{100}$ **8.** $\frac{1}{2}$ **10.** $\frac{5}{6}$ **12.** $\frac{125}{1}$ **14.** $\frac{2}{7}$ **16.** $\frac{3}{8}$

18. $T_1 = 8$ **20.** $p = \frac{105}{4}$ **22.** $R = \frac{9}{2}$ **24.** $z = .7$ **26.** $e = \frac{54}{11}$ **28.** 75.5 g

30. 0.0546 ohms   **32.** 9%   **34.** $48.86   **36.** $10\frac{1}{2}$ ft   **38.** 88   **40.** 4267

42. 7.58%   **44.** x = 24   **46.** $\frac{1}{32}$ **48.** 4.5 ohms   **50.** $E = \frac{V_3 R_t}{R_3} = 0.63$ V

52. 19.29 ft by 13.5 ft   **54.** 0.11 ft

Section 4.8

2. $T = k/p$ **4.** $y = k/x^3$ **6.** $a = kbc/d^2$ **8.** $F = km_1 m_2/r^2$ **10.** $p = kw/t$

12. $k = \frac{1}{3}$ **14.** k = 44   **16.** R = 10   **18.** $k = \frac{2}{3}$ **20.** y = 2   **22.** $y = \frac{1}{3}$

24. y = 90   **26.** 40,000   **28.** $\frac{4}{9}$ **30.** $16,085.42   **32.** 10.9   **34.** 1193.18

36. 31 in.   **38.** 121.3   **40.** 579.6 ft   **42.** 55.6 hp   **44a.** 93.3, 0.429

44b. 300, 0.667   **44c.** 14, 0.5   **44d.** 1200, 1.5   **44e.** 180, 3   **44f.** 52.8, 0.45

CHAPTER 5

FACTORING

Introduction

Purpose

- To recognize the number of terms in a given polynomial.

- To develop skills in four types of factoring used to factor polynomials: (1) factoring out the greatest common factor; (2) factoring trinomials; (3) factoring the difference of two squares; (4) factoring perfect square trinomials.

- To use factoring as a tool to solve quadratic equations.

Chapter 5 has six sections. They are: 5.1 The Greatest Common Factor; 5.2 The Product of Two Binomials; 5.3 Factoring Trinomials of the Form $x^2 + bx + c$; 5.4 Factoring Trinomials of the Form $ax^2 + bx + c$; 5.6 Solving Equations by Factoring. The six sections are followed by 66 review problems and a Self-Evaluation Test.

Chapter Objectives

(5.1) (1) Factor polynomials by using the greatest common factor.
(5.2) (2) Multiply two binomials.
(5.3) (3) Factor trinomials of the form $x^2 + bx + c$.
(5.4) (4) Factor trinomials of the form $ax^2 + bx + c$.
(5.5) (5) Square a binomial.
       (6) Factor a perfect square trinomial.
       (7) Factor the difference of two squares.
(5.6) (8) Solve quadratic equations by factoring.

Lecture Notes and Suggestions

- Have students learn the types of factoring used on polynomials:

| binomial
(two terms) | 1. G.C.F.
2. Difference of two squares |

| trinomial
(three terms) | 1. G.C.F.
2. $x^2 + bx + c$ or $ax^2 + bx + c$
3. Perfect square trinomial |

| polynomial
(four terms or more) | 1. G.C.F. |

Suggested Assignments

- Work even-numbered problems in class.

- Assign one problem from each section of the Chapter Review. For each wrong answer, assign the other review problems in the section.

- Assign odd-numbered exercises.

Section 5.1

2. $3(T_r - 2)$ **4.** $P(D - W)$ **6.** $2Y(4Y + 1)$ **8.** $6a(1 - 7a)$ **10.** $3.14(r_1 + 5r_2)$

12. $\frac{1}{12}X^3Y(2X^2 - 9Y^2)$ **14.** $RS^2(100S - R)$ **16.** $-2(3f - 2f_sk)$ **18.** $17(a + 2b - 3c)$

20. $7(V_1 + 2V_2 - 4V)$ **22.** $g^2(4 - 5g + 6g^2)$ **24.** $at(L + L_0 - Lt)$

26. $P_1(V_1 + V_2 - 2)$ **28.** $5y(a - 5b + 3c)$ **30.** $3r^2(\pi - 3\pi h - 1)$

32. $7a^2b^3(3ab - 13a + 6b^2)$ **34.** $\frac{1}{8}(4f^2 - 6F - 5)$ **36.** $0.2(T_0 - 2T_1 + 30T_2)$

38. $2.7t^3(3t^3 - 10t + 2)$ **40.** $5X^2(5X^6 - 3X^3 - 2X + 7)$ **42.** $(R - 3)(R_1 + R_2)$

44. $(T_1 + T_2)(2T_0 - 3)$ **46.** $(p + q)(1 - r)$ **48.** $P = 2(1 + w)$

50. $V = 3\pi h(a^2 + b^2 + 2h^2)$ **52.** $V = 3(x^2 + 10x + 25)$ **54.** $h = 18t(10 - t)$

56. $4.15Y^3(2Y^2 + 1)$ **58.** $19RT^2(RT - 5R + 17T)$

Section 5.2

2. $y^2 + 9y + 14$ **4.** $25b^2 + 20b + 3$ **6.** $m^2 - 8m + 7$ **8.** $20 - 16t + 3t^2$

10. $z^2 - 6z - 55$ **12.** $10 + 3n - n^2$ **14.** $25x^2 + 30xy + 8y^2$ **16.** $4r^2 + 20rt - 11t^2$

18. $9/16x^2 - 3/10x + 1/25$ **20.** $144Q^2 - 24Q - 35$ **22.** $121D^2 + 44DE - 21E^2$

24. $40c^2d^2 - 47cde + 12e^2$ **26.** $x^2 + 5x$ **28.** $S^2 - 9$ **30.** $a^2 - 36$

32. $x^2 - 1$ **34.** $4a^2 - 25$ **36.** $25b^2 - 36$ **38.** $4R^2 - 25S^2$ **40.** $81d^2 - 49e^2$

42. $0.09p^2 - q^2$ **44.** $\frac{9}{25}D^2 - E^2$ **46.** $\frac{1}{9}r^2 - \frac{1}{25}t^2$ **48.** 399

50. $3.68 S^2 + 4.65 S - 0.92$ **52.** $0.0432x^2 + 0.018x - 1.8$

Section 5.3

2. $(y + 5)(y + 1)$ **4.** $(a + 4)(a + 5)$ **6.** $(z - 8)(z - 2)$ **8.** $(R - 5)(R - 2)$

10. $(W - 7)(W + 2)$ **12.** $(B - 2)(B + 5)$ **14.** $(y - 11)(y + 1)$ **16.** $(q - 2)(q - 2)$

18. $(F + 2)(F + 4)$ **20.** $(X - 2)(X + 13)$ **22.** $(X + 7Y)(X + Y)$

24. $(m - 13n)(m - n)$ **26.** $(c + 9d)(c - d)$ **28.** $(ab + 4)(ab + 1)$

30. $4(Y - 5)(Y - 1)$ **32.** $9(F - 3)(F + 1)$ **34.** $8(T - 2)(T + 1)$

36. $a^2(b - 4)(b - 5)$ **38.** $(X + 11)(X - 5)$ **40.** $(T - 9)(T - 11)$

42. $(I - 3)(I - 7)$ **44.** $(Y - 18)(Y - 91)$ **46.** $324(j - 2)(j - 1)$

Section 5.4

2. $(2Y - 11)(Y + 1)$ 4. $(3b - 5)(5b - 1)$ 6. $(5f - 1)(f - 3)$ 8. $(d + 1)(2d - 3)$

10. $(3t + 5)(2t - 1)$ 12. $(4x - 3)(x - 2)$ 14. $(9K + 2)(K - 4)$

16. $(2p + 3)(2p + 5)$ 18. $(3a - b)(a - 2b)$ 20. $(3t + 2u)(6t + 7u)$

22. $(2X - 5)(2X + 3)$ 24. $(2A - 3)(6A - 7)$ 26. $2(2X - 7)(X - 1)$

28. $(2Y - 11)(Y - 1)$ 30. $5F(3E - 5)(5E + 1)$ 32. $2(3i - 7)(2i - 5)$

34. $(2d - 9)(2d - 9)$ 36. $3(3L - 7)(L - 1)$

Section 5.5

2. $m^2 - 4m + 4$ 4. $e^2 + 12e + 36$ 6. $16N^2 - 8N + 1$ 8. $9h^2 + 6h + 1$

10. $49r^2 - 28rs + 4s^2$ 12. $64 + 16d + d^2$ 14. $a^2 - 2ab + b^2$

16. $0.215x^2 + 1.376x + 2.2016$ 18. $(m - 8)^2$ 20. $(b + 1)^2$ 22. $(6 + n)^2$

24. $(3a - 1)^2$ 26. $(7y + 3)^2$ 28. $(9x - 7y)^2$ 30. $3(F + 11)^2$ 32. $(r - s)^2$

34. $(4W - 5U)^2$ 36. $(2b + 9c)^2$ 38. $2x + 13$ 40. $(x - 1)^2 + (y - 5)^2 = 16$

42. $(d - 8)(d + 8)$ 44. $(m + 10)(m - 10)$ 46. $(7 - r)(7 + r)$

48. $(2t - 5)(2t + 5)$ 50. $(5x + 12)(5x - 12)$ 52. $(7m - 6n)(7m + 6n)$

54. $(1/10a - b)(1/10a + b)$ 56. $(3/5S - 2/7)(3/5S + 2/7)$ 58. $7(Y + 5)(Y - 5)$

60. $(4x^2 + 9)(2x + 3)(2x - 3)$ 62. $4(x + 1)$ 64. $(Y + 68)(Y - 68)$

66. $(17j - 31)^2$

Section 5.6

2. $W = -2$ or $W = -5$ 4. $d = 2$ or $d = -9$ 6. $b = 2$ 8. $e = 3$ or $e = -4$

10. $Y = 0$ or $Y = \frac{1}{2}$ 12. $p = 0$ or $p = 9$ 14. $h = -\frac{2}{5}$ or $h = -\frac{7}{6}$ 18. $P = -\frac{3}{4}$

or $P = 5$ 20. $m = \frac{4}{3}$ or $m = -\frac{3}{4}$ 22. $T = 2$ or $T = \frac{1}{3}$ 24. 2 cm   26. 3

28. 6, 7   30. 1m   32. 7.5 sec   34. 25

ALGEBRAIC FRACTIONS

Introduction

Purpose

- To reinforce basic fraction concepts from arithmetic.

- To extend basic fraction concepts to algebraic fractions.

- To illustrate the ways in which algebraic fractions relate to technical problems.

Chapter 6 includes five sections. They are: 6.1 Algebraic Fractions in Lowest Terms; 6.2 Multiplication and Division of Algebraic Fractions; 6.3 Lowest Common Denominator of Algebraic Fractions; 6.4 Addition and Subtraction of Algebraic Fractions; 6.5 Fractional Equations.
The six sections are followed by 22 review problems and a Self-Evaluation Test.

What Makes This Book Special

- A slow and gradual treatment of algebraic fractions with numerous illustrative examples that are easy for the student to follow.

Chapter Objectives

(6.1)	(1)	Reduce fractions to lowest terms.
(6.2)	(2)	Multiply algebraic fractions.
	(3)	Divide algebraic fractions.
(6.3)	(4)	Find the lowest common denominator of two or more algebraic fractions and convert them to equivalent fractions with common denominators.
(6.4)	(5)	Add algebraic fractions.
	(6)	Subtract algebraic fractions.
(6.5)	(7)	Solve fractional equations.

Lecture Notes and Suggestions

- Have the students review Sections 1.5 through 1.10 from Chapter 1.

- Have the students review factoring in Chapter 5.

Suggested Assignments

- Use the even-numbered exercises to reinforce the lecture.

- Assign all odd-numbered exercises.

- Assign all word problems.

Section 6.1

2. $\dfrac{a - 2}{5}$ 4. $\dfrac{2}{3}$ 6. $\dfrac{u}{v + w}$ 8. $\dfrac{A + R}{A - R}$ 10. $\dfrac{5}{a + 3}$ 12. $\dfrac{1}{D - 3}$ 14. $\dfrac{x - 3y}{2}$

16. $\dfrac{N + 5}{N + 3}$ 18. $\dfrac{p + 2}{p + 3}$ 20. $\dfrac{3R + r}{R - 5r}$ 22. $\dfrac{5}{a - b}$ 24. $\dfrac{z - 1}{z + 2}$ 26. $\dfrac{w(\pi - 4)}{\pi + 5}$

28. $3(F_x + F_y)$

Section 6.2

2. $\dfrac{x}{6}$ 4. $\dfrac{25r_1^2}{r_3^2}$ 6. $\dfrac{2x^2}{5z^2}$ 8. $\dfrac{RIi^2}{3r^2}$ 10. $3(x + y)^2$ 12. $\dfrac{a(a - 1)}{b^2(a + 2)}$

14. $\dfrac{2(d - 4)}{(d - 3)(d - 1)}$ 16. $\dfrac{(g + 3h)(g + 4h)(g - h)}{(2g - h)(g + 2h)(g - 2h)}$ 18. $\dfrac{4j - 1}{(j + 3)(j + 4)}$

20. $\dfrac{9}{5(2k - 1)}$ 22. $2a$ 24. $\dfrac{18}{a^2A}$ 26. $\dfrac{2B(B + 2)}{b^2}$ 28. $\dfrac{v(w - 3)}{10w}$

30. $\dfrac{5(M + 2N)}{3(M + 4N)}$ 32. $(\theta + 5)(4\theta + 5)$ 34. $\dfrac{w - 1}{w(w + 2)(w + 1)}$ 36. $\dfrac{(2r - s)(r - 3s)}{2(r + 3s)(r - s)}$

38. $\dfrac{(x - 1)^2}{(2x - 1)^2}$ 40. $\dfrac{3n}{2(m - n)(m + n)}$ 42. $\dfrac{2(R + 1)}{R}$

Section 6.3

2. $\dfrac{25y^2}{90x^3y}$, $\dfrac{21x^2}{90x^3y}$ 4. $\dfrac{12\beta}{12\alpha^2\beta^4}$, $\dfrac{8\alpha\beta^3}{12\alpha^2\beta^4}$, $\dfrac{9}{12\alpha^2\beta^4}$ 6. $\dfrac{h^2 + 2h}{(h - 2)(h - 3)(h + 2)}$,

$\dfrac{2h^2 - 7h + 3}{(h - 2)(h - 3)(h + 2)}$ 8. $\dfrac{Z - 1}{Z(Z + 1)(Z - 1)}$, $\dfrac{Z}{Z(Z + 1)(Z - 1)}$ 10. $\dfrac{a^2b}{3a(a - 2)}$, $\dfrac{6a}{3a(a - 2)}$

12. $\dfrac{x^3 - 4x^2 + 4x}{(x + 3)(x - 3)(x - 2)^2}$, $\dfrac{2x^2 - 10x + 12}{(x + 3)(x - 3)(x - 2)^2}$, $\dfrac{x^3 - 9x}{(x + 3)(x - 3)(x - 2)^2}$

14. $\dfrac{\theta^2 + \theta\varnothing}{(\theta + \varnothing)(\theta - \varnothing)}$, $\dfrac{\theta\varnothing - \varnothing^2}{(\theta + \varnothing)(\theta - \varnothing)}$ 16. $\dfrac{10Z - 10}{10(Z - 5)(Z + 5)}$, $\dfrac{2Z^2 + 10Z}{10(Z - 5)(Z + 5)}$,

$\dfrac{Z^2 - 7Z + 10}{(10(Z - 5)(Z + 5)}$ 18. $\dfrac{3r - 9}{(r + 2)(r + 1)(r - 3)}$, $\dfrac{2r + 4}{(r + 2)(r + 1)(r - 3)}$

Section 6.4

2. $\dfrac{9r - 2s}{105}$ 4. $\dfrac{3R - h}{12R^3}$ 6. $\dfrac{16N + 25M^2}{60M^3N^2}$ 8. $\dfrac{4\lambda - \lambda^2}{\lambda - 1}$ 10. $\dfrac{2x^2 - 5x - 13}{(x - 2)(x - 5)}$

12. $\dfrac{d + 5L}{(d - L)(d + L)}$ 14. $\dfrac{I^2 - 4I}{(I - 1)(I + 1)}$ 16. $\dfrac{-E^2 + 9E - 3}{(E - 3)(E - 2)(E + 2)}$

18. $\dfrac{\beta(\alpha^2 - 6)}{3\alpha(\alpha - 2)}$ 20. $\dfrac{-5f^2 + 24f - 8}{(f - 5)(f - 4)}$ 22. $\dfrac{-w^2 - 7w}{10(w + 5)(w - 5)}$ 24. $\dfrac{GMm - mrv^2}{r^2}$

26. $\dfrac{C_2C_3 + C_1C_3 + C_1C_2}{C_1C_2C_3}$ **28.** $\dfrac{f\ell^4 + 8F\ell^3}{384\ EI}$

Section 6.5

2. $a = \dfrac{6}{13}$ **4.** $G = 6$ **6.** $W = \dfrac{8}{3}$ **8.** $x = \dfrac{7}{4}$ **10.** $A = -\dfrac{16}{5}$ **12.** $e = \dfrac{13}{5}$

14. $x = \dfrac{25}{13}$ **16.** $T = \dfrac{23}{3}$ **18.** $A = -\dfrac{30}{17}$ **20.** $a = \dfrac{3}{2}$ **22.** $z = \dfrac{4}{5}$

24. all reals except 1, −1, and −4   **26.** no solution   **28.** $i = \dfrac{nE + IRe}{InR}$

30. $j = 2$ **32.** $C = \dfrac{C_1C_2}{C_1 + C_2}$ **34.** $L = \dfrac{P}{pD}$ **36.** $g = \dfrac{WV^2}{2KE}$ **38.** $r = \dfrac{mv^2}{F-mg}$

40. $x = 6m$ **42.** $\dfrac{3}{4}, \dfrac{3}{2}$ **44.** 1

CHAPTER 7

GRAPHS

Introduction

Purpose

- To express data using circle graphs, bar graphs, and line graphs.

- To determine if a given relation is a function, and to find the domain and range of a given function.

- To graph straight lines by graphing points on the line or by using the slope-intercept method.

Chapter 7 includes four sections. They are: 7.1 Graphic Methods for Presenting Collected Data; 7.2 The Rectangular Coordinate System; 7.3 Graphing Linear Equations in Two Variables; 7.4 The Slope-Intercept Form of an Equation of a Straight Line. The four sections are followed by 46 review problems and a Self-Evaluation Test.

What Makes This Book Special

- In Chapter 2, students learned to recognize geometric figures and to find their perimeter and area. These figures are again used in Chapter 7 to teach students to plot points in the rectangular coordinate system.

- The concept of graphing data using circle graphs, bar graphs, and line graphs is extended to graphing data in the plane.

Chapter Objectives

(7.1)   (1) Express given data using circle graphs, bar graphs, and line graphs.
(7.2)   (2) Plot points in the rectangular coordinate system.
(7.3)   (3) Graph a linear equation in two variables.
(7.4)   (4) Graph a linear equation using the slope-intercept method.

Lectures Notes and Suggestions

- Have students collect data from their own experiences, and illustrate this data by using a circle graph, a bar graph, or a line graph.

- Discuss why a particular type of graph was used.

- Graph an equation such as $y = \frac{2}{3} x - 6$ by using points (0, -6), (3, -4), (6, -2), (9, 0), and (12, 2) to illustrate the meaning of the slope $\frac{2}{3}$. As x increases 3, y increases 2. Have the students make up their own examples to illustrate this concept.

Suggested Assignments

- Use the problems in Exercise 7.1 to draw all three types of graphs wherever possible.

- Graph the equations in Exercise 7.3 using the slope-intercept method if more practice is needed.

25

Section 7.1

2.

4. Full charge
    1/4 discharge
    1/2 discharge
    3/4 discharge
    Discharge

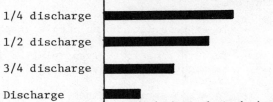

6.

8.

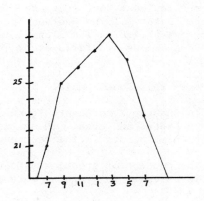

10.

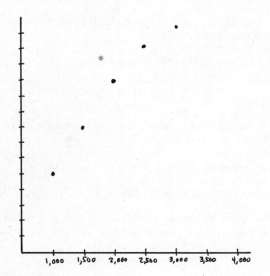

2.
4.
6.
8.
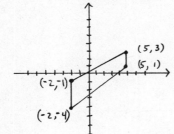

10.
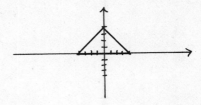

12.
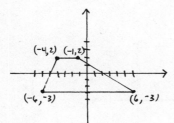

14.
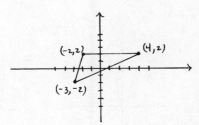

16.
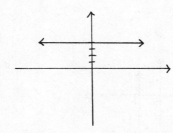

18.
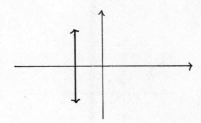

20.

22. I and IV   24. I and II   26a. (−5, 1)   26b. (−3, 0)   26c. (−1/2, 1/2)

26d. (0, 5)   26e. (1.5, 4)   26f. (4 1/2, 1 1/2)   26g. (2, 0)   26h. (7, −1)

26i. (4, −3)   26j. (0, −5)   26k. (−1, −4)   26ℓ. (−5, −3)

28. D = {0, 1, 2, 3, 4}   30. D = {−5, −1, 3}   32. D = {1}
    R = {1}   R = {−5, −1, 3}   R = {2, 3, 4}
    Function   Function   Not a function

27

34. 37.5 sq in.

36.

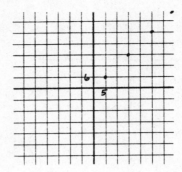

2.

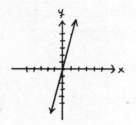

4.

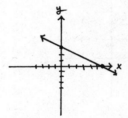

6.

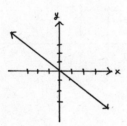

8.

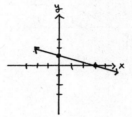

10.

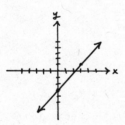

12.

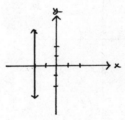

14.

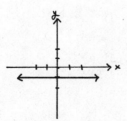

16.

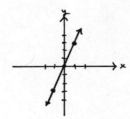

18.

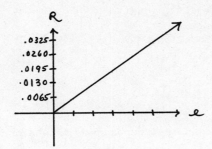

2. $\frac{3}{4}$, 9   4. $\frac{1}{3}$, -2   6. $\frac{4}{5}$, -3   8. 3   10. $-\frac{3}{7}$ 12. $\frac{3}{2}$

14.

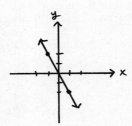

16.

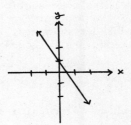

18.

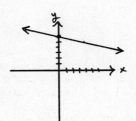

20.

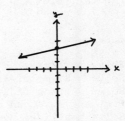

22.

24.

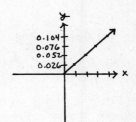

29

CHAPTER 8

SYSTEMS OF LINEAR EQUATIONS

Introduction

Purpose

- Determine if a system of two linear equations is independent, inconsistent, or dependent.

- To solve a system of two linear equations by (1) graphing; (2) addition; (3) substitution; and (4) Cramer's Rule.

- To solve a system of three linear equations by Cramer's Rule.

Chapter 8 includes four sections. They are: 8.1 Solving Linear Equations by Graphing; 8.2 Solving Systems of Linear Equations Algebraically; 8.3 Second Order Determinants; 8.4 Third Order Determinants.
The four sections are followed by 40 review problems and a Self-Evaluation Test.

What Makes This Book Special

- The methods of graphing a straight line from Chapter 7 are reinforced in Chapter 8.

- The types of word problems presented in Chapter 4 are presented again to be solved using two variables and two equations.

Chapter Objectives

(8.1)   (1) Solve systems of linear equations in two variables by graphing.
(8.2)   (2) Solve systems of linear equations in two variables by the addition method.
        (3) Solve systems of linear equations in two variables by the substitution method.
(8.3)   (4) Evaluate a second order determinant.
        (5) Solve two linear equations in two variables by using Cramer's Rule.
(8.4)   (6) Evaluate a third order determinant.
        (7) Solve three linear equations in three variables by using Cramer's Rule.

Lecture Notes and Suggestions

- Use graphing to teach the student to understand the vocabulary: independent, inconsistent, and consistent.

- From solving a system of linear equations by graphing, progress to algebraic methods and Cramer's Rule.

- Solve problems like problem 36 in Exercise 8.2 using one variable and one equation. Have students compare the two methods.

Suggested Assignments

- Have student solve a system of two linear equations by graphing, addition, substitution, and by using Cramer's Rule. Discuss which method is preferred, and why it is preferred.

- Check the methods used by students in Exercise 8.2 when the instructions are to solve by any method.

Section 8.1

2. (-2, 5)   **4.** (0, -1)   **6.** Dependent   **8.** (1, 1)   **10.** Dependent

12. (-2, -4)   **14.** t = -3, h = -6   **16.** Inconsistent   **18.** ℓ = 4m, w = 1 m

20. 2 in., 4 in.

Section 8.2

2. (6, -2)   **4.** (0, 3)   **6.** (8/5, 19/20)   **8.** (-6, 4)   **10.** (3, 0)

12. Dependent   **14.** (28/29, -50/29)   **16.** (-2, 3)   **18.** (12/19, -13/38)

20. Dependent   **22.** (-25/19, -1) or (-1.3, -1)   **24.** (11/6, -1/4)

26. Inconsistent   **28.** (-3/2, -9/2)   **30.** (32/5, -7/5)   **32.** (13/5, -4/5)

34. Inconsistent   **36.** 33 and 54   **38.** 3 cm by 11 cm   **40.** 70 kg and 110 kg

42. 40 size H, 32 size G   **44.** 400 kg, 600 kg   **46.** W_1 = 8 lb, W_2 = 12 lb

48. 36.8 ft/min, 48.2 ft/min

Section 8.3

2. 1   **4.** 4   **6.** 0   **8.** (3, 2)   **10.** Inconsistent   **12.** (-1, 4)

14. (3/25, 4/25)   **16.** (9/11, 7/11)   **18.** (.3, -.4)   **20.** (6, 12)

22. Dependent   **24.** (2, -2)   **26.** (-3, 4)   **28.** Inconsistent   **30.** 9 cm by 19 cm

32. $8,500 at 11%, $3,500 at 13%   **34.** $44,400   **36.** 20°, 70°

38. 15ℓ **40.** 6.5 m, 2.3 m

Section 8.4

2. 1   **4.** 0   **6.** 259   **8.** 0   **10.** .16524   **12.** (-1, 0, 6)

14. (3/5, -1/5, 2/5)   **16.** no solution   **18.** (1/2, -5/2, -7/2)   **20.** (5, -6, 2)

22. (1/2, 3/4, -1/2)   **24.** 400 kg, 600 kg, 1,000 kg   **26.** 14 size G, 16 size H,

and 42 size K

EXPONENTS AND SCIENTIFIC NOTATION

Introduction

Purpose

- To extend the concepts of exponents and scientific notation introduced in Chapter 1.

- To relate the concepts discussed to technical problems.

Chapter 9 includes six sections. They are: 9.1 Positive Exponents; 9.2 Zero Exponents; 9.3 Negative Exponents; 9.4 Fractional Exponents; 9.5 Scientific Notation; 9.6 Calculations in Scientific Notation.
The six sections are followed by 45 review problems and a Self-Evaluation Test.

What Makes This Book Special

- A straightforward discussion of the concepts, followed by numerous illustrative examples.

Chapter Objectives

(9.1)   (1) Simplify expressions with positive integral exponents.
(9.2)   (2) Simplify zero exponents.
(9.3)   (3) Simplify negative exponents.
(9.4)   (4) Simplify fractional exponents.
(9.5)   (5) Express a given number in scientific notation.
(9.6)   (6) Multiply numbers expressed in scientific notation.
(9.7)   (7) Divide numbers expressed in scientific notation.

Lecture Notes and Suggestions

- Have the students review Sections 1.3 and 1.13 from Chapter 1 that relate to exponents and scientific notation.

Suggested Assignments

- Use the even-numbered exercises to reinforce the lecture.

- Assign all odd-numbered exercises.

- Assign all word problems.

Section 9.1

2. a^6 4. 10^{11} 6. 5^8 8. y^{15} 10. π^2 12. 10 14. -5 16. $\dfrac{1}{t^2}$

18. $\dfrac{1}{10^4}$ 20. $\dfrac{1}{d^8}$ 22. $r_1^4 r_2^4$ 24. $125 i^3 r^3$ 26. $\dfrac{v_0^7}{v_f^7}$ 28. $\dfrac{81}{z_a^2}$

30. g^9 32. 10^{16} 34. 3^{12} 36. $16 p^{16}$ 38. $u^6 v^{12} w^3$ 40. $\dfrac{i^6}{16}$

42. $\dfrac{1}{81 \psi^4}$ 44. $\dfrac{1}{10^8}$ 46. $R^8 s^5$ 48. $\dfrac{v}{v_0}$ 50. $\dfrac{1}{10^3 d}$ 52. $5^{11} f^{23} h^{18} j^3$

54. $\dfrac{49 y^4}{729 \mu^6}$ 56. $7^3 10^{15}$ 58. $36 \pi r^6$

Section 9.2

2. 1 4. 1 6. 1 8. 5 10. 1 12. 1 14. $\dfrac{1}{27 z_3^3}$ 16. 10

18. $\dfrac{1}{81 \phi^4}$ 20. $\alpha^3 \gamma^3$ 22. 1 24. $w - 2$ 26. 1

Section 9.3

2. $\dfrac{1}{H^5}$ 4. $\dfrac{1}{9}$ 6. b 8. 75 10. $\dfrac{-2}{j^3}$ 12. $\dfrac{1}{9 s^2}$ 14. $\dfrac{k}{3}$ 16. $-125 p^3$

18. $\dfrac{1}{10}$ 20. $\dfrac{1}{16}$ 22. $\dfrac{1}{Q^5}$ 24. 10^6 26. $\dfrac{1}{P Q^2 R}$ 28. $\dfrac{1}{10^5}$ 30. $\dfrac{9 j^3}{h}$

32. $\dfrac{1}{125 y^9}$ 34. $\dfrac{R_i^{15}}{R_0^{10}}$ 36. $\dfrac{m^6}{L^4 d^2}$ 38. $\dfrac{49 s^2}{360^6 n^4}$ 40. $5 s$ 42. $\dfrac{4}{9 R^9}$

44. $\dfrac{3}{A^3}$ 46. $\dfrac{36 \gamma^7}{\mu^{15}}$ 48. $t_0 - \dfrac{1}{t^2}$ 50. $5 - \dfrac{2}{e^3}$ 52. $W = 2 m \pi^2 f^2 A^2$

54. $R = \dfrac{R_a R_f}{R_a + R_f}$ 56. $R_1 = \dfrac{I_1^2 m_2 R_2}{\ell_2^2 m_1}$ 58. $\dfrac{1}{14,348,907 x^{20}}$

2. $B^{3/2}$ **4.** $K^{5/24}$ **6.** $\dfrac{1}{W^{1/4}}$ **8.** $e^{1/2}i^{1/2}$ **10.** $\dfrac{r^{1/3}}{3^{1/3}}$ **12.** $c^{1/8}$

14. $\dfrac{Q^{1/3}}{P^{7/4}}$ **16.** $\dfrac{E}{I^{1/8}}$ **18.** $R^{2/5}C^{2/5}$ **20.** $\dfrac{V_c(17)^{1/2}}{2}$

Section 9.5

2. 3.05×10^2 **4.** 4.5×10^3 **6.** 4.72×10^{-1} **8.** 2.985×10^2

10. 7.9×10^{-3} **12.** 5.4×10^7 **14.** 6.8×10^{-5} **16.** $90,000,000$

18. 0.000000081 **20.** 0.00002 **22.** $8.13 \times 10^{-3}, 3$ **24.** $7.885 \times 10^5, 4$

26. $8 \times 10^5, 1$ **28.** $3.06 \times 10^{-2}, 3$ **30.** $18,000,000$ **32.** 6×10^3

34. 2.8×10^{-4}

Section 9.6

2. 1.36×10^{12} **4.** 8.8×10^{12} **6.** 5×10^3 **8.** $0.016\overline{6}$ **10.** 4.6×10^{-6}

12. 5.36 **14.** 6.91478×10^4 **16.** 1.0063×10^{-4}

CHAPTER 10

ROOTS AND RADICALS

Introduction

Purpose

- To develop manipulative skills for solving problems involving roots and radicals.

- To illustrate the ways in which roots and radicals are important to the technical student.

Chapter 10 includes eight sections. They are: 10.1 Roots of Numbers; 10.2 Radical Expressions and Exponents; 10.3 Simplifying Radicals; 10.4 Multiplication of Radicals; 10.5 Division of Radicals; 10.6 Addition and Subtraction of Radicals; 10.7 Multiplication of Binomials Containing Radicals; 10.8 Rationalizing Binomial Denominators. The eight sections are followed by 51 review problems and a Self-Evaluation Test.

What Makes This Book Special

- Numerous word problems from technology involving roots and radicals.

- Several straightforward examples.

Lecture Notes and Suggestions

- Encourage students to use calculators wherever possible to solve problems.

Suggested Assignments

- Use the even-numbered exercises to reinforce the lecture.

- Assign all even-numbered exercises.

- Assign all word problems.

Section 10.1

2. 3   4. 5   6. 2   8. -1   10. -2   12. 10   14. no real root

16. 1   18. 12   20. -4   22. -9   24. 2   26. 0.2   28. 40   30. 0.5

32. 0.13   34. 14

Section 10.2

2. m^2 4. f   6. e   8. a   10. -x   12. $-B^3$ 14. r^4 16. t^4

18. $5r^2$ 20. 2a   22. xy^2z 24. $2W^4$ 26. $d^{1/3}$ 28. $a^{5/4}$ 30. $(mp)^{1/7}$

32. $\sqrt[5]{t}$ 34. $\sqrt[4]{st}$ 36. $\sqrt[7]{25}$ 38. $\sqrt[4]{n^3}$ 40. $\sqrt{7}$ 42. R   44. 2

46. 4   48. 8   50. $\frac{1}{4}$ 52. $\frac{1}{8}$ 54. 3.46   56. 6.91

Section 10.3

2. $3\sqrt{3}$ 4. $2\sqrt{7}$ 6. $10\sqrt{2}$ 8. $6\sqrt{3}$ 10. $-5\sqrt{5}$ 12. $15\sqrt{2}$ 14. $r\sqrt{r}$

16. $-y^5\sqrt{y}$ 18. $rt^2\sqrt{rt}$ 20. $2d^4\sqrt{6d}$ 22. $-5\sqrt{41}$ 24. $\sqrt{x}$ 26. $5\sqrt[3]{3}$

28. $2\sqrt[4]{2}$ 30. $4M^2\sqrt[3]{4M}$ 32. $-7AB\sqrt{2BC}$ 34. $\sqrt{m}$ 36. -3   38. $T\sqrt{W}$

40. 75   42. 60   44. 1.3

Section 10.4

2. $\sqrt{30}$ 4. $\sqrt{39}$ 6. 9   8. $\sqrt[4]{12}$ 10. $5\sqrt{6}$ 12. $x\sqrt{15}$ 14. $y\sqrt{35}$

16. $\sqrt[5]{36m^4}$ 18. $4uv\sqrt{7u}$ 20. $-6\sqrt{33}$ 22. $54\sqrt{2}$ 24. $a^2b^2\sqrt[4]{b}$

26. $4\sqrt{15}$ 28. $-5\sqrt{6}$ 30. $18r^2s\sqrt{rs}$ 32. $-30CD\sqrt[3]{2C^2D}$ 34. $\sqrt{10}+\sqrt{35}$

36. $-66 - 30\sqrt{66} + 6\sqrt{110}$

Section 10.5

2. $\sqrt{5}$ 4. $\sqrt{3}$ 6. $\sqrt[3]{9}$ 8. x   10. $6\sqrt{P}$ 12. $\sqrt[3]{9}$ 14. $\dfrac{-\sqrt{15}}{5}$

16. $\dfrac{-\sqrt{66}}{11}$ 18. $\dfrac{\sqrt{2y^2}}{y}$ 20. $\dfrac{\sqrt{st}}{t^2}$ 22. $\dfrac{4\sqrt{15}}{3}$ 24. $\dfrac{\sqrt{15cd}}{3c}$ 26. $\dfrac{3\sqrt{t}}{t^2}$

28. $\dfrac{x\sqrt{y}}{2y}$ 30. $\dfrac{\sqrt[4]{8Z^3}}{Z}$ 32. $\dfrac{\sqrt[4]{4\pi^3}}{\pi}$

Section 10.6

2. $\sqrt{11}$ **4.** $2\sqrt[4]{3r}$ **6.** $-4\sqrt{5}$ **8.** $12 - 12\sqrt{2}$ **10.** $-\sqrt{3q}$ **12.** $-2t\sqrt{t}$

14. $\sqrt{15}$ **16.** $-0.3\sqrt{m}$ **18.** $4\sqrt{10}$ **20.** $2\sqrt{7} - 4\sqrt{11}$ **22.** $4\sqrt[4]{xy}$

Section 10.7

2. $31 - 11\sqrt{3}$ **4.** $-11 + 13\sqrt{7}$ **6.** -16 **8.** $55x - 2\sqrt{x} - 21$ **10.** $17 - 12\sqrt{2}$

12. $36 - 5\sqrt{T} - T$ **14.** $a - 2b\sqrt{a} + b^2$ **16.** $W^2 + 2W\sqrt{6} - 18$ **18.** $6\sqrt{35} + 596$

20. $18 - 7\sqrt{P} - \sqrt[3]{p^2}$

Section 10.8

2. $\dfrac{11(\sqrt{5} - 3)}{-4}$ **4.** $\dfrac{12 + \sqrt{11}}{133}$ **6.** $\dfrac{5\sqrt{3} + \sqrt{6}}{23}$ **8.** $\dfrac{11 + 5\sqrt{5}}{4}$

10. $\dfrac{32 - 5\sqrt{10}}{18}$ **12.** $\dfrac{6 + 5\sqrt{f} + f}{4 - f}$ **14.** $\dfrac{M + 8\sqrt{M} + 12}{M - 4}$ **16.** $3\sqrt{7} - 8$

18. $\dfrac{X - 2\sqrt{XY} + Y}{X - Y}$ **20.** $\dfrac{82 + 2\sqrt{6}}{67}$

QUADRATIC EQUATIONS

Introduction

Purpose

- To solve quadratic equations that cannot be solved by factoring.

- To add, subtract, multiply, and divide complex numbers.

- To solve quadratic equations that have complex solutions.

Chapter 11 has six sections. They are: 11.1 Solving Quadratic Equations of the Form $x^2 = a$; 11.2 Solving Quadratic Equations by Completing the Square; 11.3 The Quadratic Formula; 11.4 Complex Numbers; 11.5 Multiplication and Division of Complex Numbers; 11.6 Quadratic Equations With Complex Solutions.
The six sections are followed by 68 review problems and a Self-Evaluation Test.

What Makes This Book Special

- Geometric formulas from Chapter 2 are used for word problems in this chapter.

- The method of solving quadratic equations in Chapter 5 is reviewed before presenting additional methods of solving quadratic equations.

- Complex numbers are introduced in this chapter in order to be able to solve quadratic equations that have complex solutions.

Chapter Objectives

(11.1)	(1)	Solve quadratic equations by taking the square root of both members of the equation.
	(2)	Solve problems using the Pythagorean Theorem.
	(3)	Find the distance between two points.
(11.2)	(4)	Solve a quadratic equation by completing the square.
(11.3)	(5)	Solve quadratic equations using the quadratic formula.
(11.4)	(6)	Add and subtract complex numbers.
(11.5)	(7)	Multiply complex numbers.
	(8)	Divide complex numbers.
(11.6)	(9)	Solve quadratic equations that have complex solutions.

Lecture Notes and Suggestions

- Use an example, such as $x^2 + 2x - 5 = 0$, to illustrate that all quadratic equations cannot be solved by factoring.

- Use this same example to illustrate that the method of completing the square and the quadratic formula give the same solutions. Find out which method is preferred by the majority of the students.

- Have students illustrate properties of real numbers (commutative, associative, distributive, identities, and inverses) for the set of complex numbers.

Suggested Assignments

- Use even-numbered exercises in class to reinforce the lecture.

- Assign even-numbered exercises.

- Work the problems in Exercise 5.6 using the method of completing the square, and the quadratic formula.

Section 11.1

2. $d = \pm 4$ 4. $A = \pm \sqrt{5}$ 6. $y = \pm 3\sqrt{7}$ 8. No solution   10. $y = 12, -8$

12. $R = -11, 5$ 14. $p = -3 \pm 2\sqrt{2}$ 16. $R = 9 \pm 4\sqrt{3}$ 18. $e = -7 \pm \sqrt{7}$

20. $a = \dfrac{4}{5}, -\dfrac{2}{5}$ 22. $Q = \dfrac{2 \pm \sqrt{7}}{7}$ 24. $S = \dfrac{15 \pm 2\sqrt{15}}{5}$ 26. 24

28. $\sqrt{74}$ 30. 40 m   32. 3.8 km   34. 5308 miles   36. $\sqrt{65}$ m $\doteq$ 8.1 m

38. $\sqrt{13}$ 40. $\sqrt{17}$ 42. $2\sqrt{5}$ 44. 19.7   46. 10.28

Section 11.2

2. $t = 3, -9$ 4. $y = \dfrac{-1 \pm \sqrt{5}}{2}$ 6. $M = \dfrac{-3 \pm \sqrt{17}}{2}$ 8. $d = 2, -\dfrac{7}{2}$

10. $m = \dfrac{5}{2}, -3$ 12. $b = -\dfrac{5}{2}, \dfrac{1}{3}$ 14. $D = \dfrac{-3 \pm \sqrt{129}}{10}$ 16. 6 inches

18. -20 and -4, or 4 and 20

Section 11.3

2. $m = -6 \pm \sqrt{6}$ 4. $R = -1, 1/5$ 6. $i = 3, 1$ 8. $p = \dfrac{-5 \pm \sqrt{41}}{8}$

10. $b = \dfrac{-9 \pm \sqrt{21}}{10}$ 12. $v = -3/2, 1/5$ 14. $r = \dfrac{-3 \pm \sqrt{129}}{6}$ 16. $s = \dfrac{1 \pm \sqrt{111}}{10}$

18. $x = \dfrac{-13 \pm \sqrt{57}}{8}$ 20. $K = 1, -3/5$ 22. $T = \dfrac{3 \pm \sqrt{33}}{2}$ 24. $R = \dfrac{-3 \pm \sqrt{249}}{10}$

26. $-1 + 3\sqrt{3}$ cm (base)
$1 + 3\sqrt{3}$ cm (height)

28. 6 cm, 8 cm, 10 cm   30. $\dfrac{5 + 5\sqrt{10}}{9}$ 32. 0.25 sec, 1.75 sec

$\dfrac{15 + 15\sqrt{10}}{9}$

34. 17.65 cm, 28.65 cm, 33.65 cm $\dfrac{45 + 5\sqrt{10}}{9}$

Section 11.4

2. $10j$ 4. $-48j$ 6. $-j$ 8. $2j\sqrt{6}$ 10. $4 - 65j$ 12. $24 + j$ 14. 10

16. $2j$ 18. 0   20. $5j$ 22. $-26 + 8j$ 24. 1   26. $-5 - 10j$ 28. $-\dfrac{1}{5}$

30. $\dfrac{3}{2} + \dfrac{1}{6}j$ 32. $3 + j\sqrt{3}$ 34. $7 + 3j\sqrt{7}$ 36. $-4 - 3j$ 38. 0

40. $(44 - 10j)$ ohms **42.** $59 + 5j$ **44.** $113 + 53j$ **46.** $-21.82 + 4.02j$

48. $-797 + 262j$

Section 11.5

2. $21 - 3j$ **4.** $4 - 12j$ **6.** $-1 + 2j$ **8.** $-30 - 27j$ **10.** $48 + 53j$

12. $-14 + 11j$ **14.** -16 **16.** $-2 + 2j$ **18.** $-20 - 10j$ **20.** $22 + 14j$

22. $-113j$ **24.** $91 - j$ **26.** 17 **28.** 106 **30.** $\frac{21}{10} - \frac{7}{10} j$

32. $\frac{20}{17} + \frac{5}{17} j$ **34.** $-\frac{1}{2} + \frac{1}{2} j$ **36.** $\frac{5}{4} + j$ **38.** $\frac{21}{29} - \frac{20}{29} j$ **40.** $\frac{29}{34} - \frac{3}{34} j$

42. $27 + 24j$ **44.** $\sqrt{425} \doteq 20.6$ **46.** $\frac{15}{34} - \frac{25}{34} j$ **48.** $12.8592 - 96.7536 j$

50. $0.51 + 0.18 j$

Section 11.6

2. $m = \pm 6j$ **4.** $f = \pm j \sqrt{13}$ **6.** $b = \pm 5j \sqrt{3}$ **8.** $y = \frac{1}{2} \pm \frac{j \sqrt{19}}{2}$

10. $b = \frac{1}{3} \pm \frac{j \sqrt{26}}{3}$ **12.** $D = \frac{-1}{4} \pm \frac{j \sqrt{23}}{4}$ **14.** $X = \frac{-1}{9} \pm \frac{j \sqrt{35}}{9}$

16. $W = \frac{-5}{12} \pm \frac{j \sqrt{23}}{12}$ **18.** $R = \frac{5}{6} \pm \frac{j \sqrt{59}}{6}$ **20.** $i = \frac{-2}{3} \pm \frac{j \sqrt{26}}{3}$

22. $\frac{-5}{2} \pm \frac{j \sqrt{15}}{2}$ ohms **24.** $\frac{3}{2} \pm \frac{j \sqrt{47}}{2}$ **26.** $i = \frac{1}{7} \pm \frac{j \sqrt{83}}{7}$ amperes

28. $E = 2.90 \pm 3.90j$ **30.** $E = 0.22 \pm 1.43j$

CHAPTER 12

LOGARITHMS

Introduction

Purpose

- To develop the basic skills with logarithms.

- To strengthen the student's skill with the calculator in solving logarithm-related problems.

Chapter 12 includes seven sections. They are: 12.1 Exponential and Logarithmic Functions; 12.2 Logarithmic Equations; 12.3 Common Logarithms; 12.4 Antilogarithms; 12.5 Properties of Logarithms; 12.6 Computation Using Logarithms; 12.7 Natural Logarithms.
The seven sections are followed by 29 review problems and a Self-Evaluation Test.

What Makes This Book Special

- The emphasis on using the calculator as well as the table to solve problems.

Chapter Objectives

(12.1)	(1) Graph exponential functions.
	(2) Graph logarithmic functions.
	(3) Evaluate exponential functions when given selected values.
(12.2)	(4) Solve logarithmic equations.
(12.3)	(5) Use Table 2 or a calculator to find common logarithms.
(12.4)	(6) Find antilogarithms of numbers using Table 2 or a calculator.
(12.5)	(7) Use the properties of logarithms.
(12.6)	(8) Perform computations using logarithms.
(12.7)	(9) Find natural logarithms.

Lecture Notes and Suggestions

- Encourage students to perfect their ability to use the calculator to solve problems.

Suggested Assignments

- Use the even-numbered exercises to reinforce the lecture.

- Assign all even-numbered exercises.

CHAPTER 13

TRIGONOMETRY

Introduction

Purpose

- To introduce radian and degree measure in the first section in order to use it interchangeably throughout the twelve sections.

- To simplify and abbreviate the numerical computation by using the calculator in many examples.

- To find the trigonometric ratios of angles between 0° and 90° in order to work with them in solving right triangles.

- To work with the polar and rectangular form of vectors.

- To solve oblique triangles using the Law of Sines and the Law of Cosines.

- To graph variations of the sine and cosine functions, and to graph the tangent function.

- To solve linear and quadratic trigonometric equations.

Chapter 13 includes twelve sections. They are: 13.1 Angles; 13.2 Trigonometric Ratios; 13.3 Value of Trigonometric Ratios; 13.4 Elementary Right Triangle Applications; 13.5 Angles Larger Than 90 ; 13.6 Vectors; 13.7 Vectors and Complex Numbers; 13.8 The Law of Sines; 13.9 The Law of Cosines; 13.10 Graphs of the Sine, Cosine, and Tangent; 13.11 The Graphs of $y = a \sin (bx + 0)$ and $y = a \cos (bx + 0)$; 13.12 Trigonometric Equations.
The twelve sections are followed by 104 review problems and a Self-Evaluation Test.

What Makes This Book Special

- A step-by-step approach to trigonometry.

- The explanations for each section are kept brief. Concepts are reinforced with examples, followed by graded exercises.

- The geometry needed for this chapter was taught in Chapter 2 and in Chapter 7, and has been used in word problems throughout the text.

Chapter Objectives

(13.1)	(1) Determine the quadrant of a given angle and name coterminal angles.
	(2) Express degree measure of an angle as radian measure and convert radian measure to degree measure.
(13.2)	(3) Determine the trigonometric ratios of an acute angle of a right triangle.
(13.3)	(4) Find the value of trigonometric ratios of angles between 0° and 90°.
(13.4)	(5) Solve right triangles.
(13.5)	(6) Find the trigonometric ratios for angles larger than 90°.
(13.6)	(7) Express vectors in rectangular and polar form.
	(8) Solve problems using vectors.
(13.7)	(9) Add vectors graphically.
	(10) Express complex numbers in polar form.
	(11) Multiply vectors by the operator j.
	(12) Find the product and quotient of two complex numbers in polar form.
(13.8)	(13) Solve oblique triangles using the Law of Sines.
(13.9)	(14) Solve oblique triangles using the Law of Cosines.

(13.10)   (15) Graph curves of the form y = a sin bx.
            (16) Graph curves of the form y = a cos bx.
            (17) Graph curves of the form y = a tan bx.
(13.11)   (18) Graph equations of the form a sin (bx + θ).
            (19) Graph equations of the form a cos (bx + θ).
(13.12)   (20) Solve trigonometric equations.

Lecture Notes and Suggestions

- Give daily quizzes to make sure students memorize the definitions and rules in each section.

- Before teaching each section, choose concepts from the basic geometry in Chapter 2.

- Before solving trigonometric equations in Section 13.12, review the methods of solving linear equations from Chapter 4, and the methods of solving quadratic equations from Chapters 4 and 11.

Suggested Assignments

- In Exercise 13.1, find coterminal angles using radian measure.

- In Section 13.2, add to Table 13.1 by listing the trigonometric ratios of 0° and 90°.

- In Section 13.4 when finding the leg or hypotenuse of a right triangle, use a trigonometric ratio and the Pythagorean Theorem and compare the accuracy of the answers.

- After working on Section 13.5, complete Table 13.1 by adding the trigonometric ratios of 180° and 270°.

- When working Exercise 13.9, solve a triangle using only the Law of Cosines and then compare the answers obtained in using both the Law of Cosines and the Law of Sines.

- In Exercise 13.10, graph the equations on the same grid and add or subtract the two equations by adding or subtracting the coordinates.

46

Section 13.1

2. II, $500°$, $-220°$ **4.** IV, $345°$, $-375°$ **6.** III, $260°$, $-100°$ **8.** IV, $640°$, $-80°$

10. I, $200°30'$, $-380°30'$ **12.** IV, $660°50'$, $-59°10'$ **14.** $50.83°$ **16.** $287.73°$

18. $49.95°$ **20.** $16°42'$ **22.** $300°24'$ **24.** $40.2°$ **26.** $\dfrac{\pi}{3}$ **28.** $\dfrac{\pi}{5}$

30. $\dfrac{5\pi}{4}$ **32.** $\dfrac{3\pi}{10}$ **34.** $-\dfrac{2\pi}{5}$ **36.** $18°$ **38.** $112\frac{1}{2}°$ **40.** $315°$ **42.** $-30°$

42. $-90°$ **46.** $\dfrac{19\pi}{8}\,m^2 \doteq 7m^2$ **48.** 6.3 sq in.   **50.** $40.1\ cm^2$ **52.** $\dfrac{25\pi}{4}\ cm$

54. $\dfrac{3}{4}\pi\ cm$ **56.** $\dfrac{7\pi}{60}\ m$ **58.** 0.17 radians   **60.** 3.72 radians   **62.** $1.833°$

64. $113.961°$

Section 13.2

2. $\dfrac{5}{13}, \dfrac{12}{13}, \dfrac{5}{12}, \dfrac{13}{5}, \dfrac{13}{12}, \dfrac{12}{5}$ **4.** $\dfrac{a}{c}, \dfrac{b}{a}, \dfrac{c}{a}, \dfrac{b}{a}, \dfrac{c}{a}, \dfrac{b}{c}$ **6.** $\dfrac{12}{13}, \dfrac{12}{5}$ **8.** $\dfrac{3}{5}, \dfrac{5}{4}$

10. $\dfrac{5}{3}$ **12.** $\dfrac{5}{13}$ **14.** $\dfrac{9}{41}$ **16.** $3^2 + 4^2 = 5^2$ **18.** $8^2 + 15^2 = 17^2$

20. $1^2 + 1^2 = (\sqrt{2})^2$ **22.** $\dfrac{a}{c} \cdot \dfrac{c}{a} = 1;\ \dfrac{b}{c} \cdot \dfrac{c}{b} = 1;\ \dfrac{a}{b} \cdot \dfrac{b}{a} = 1$

24. $\left(\dfrac{a}{b}\right)^2 + 1 = \left(\dfrac{c}{b}\right)^2$ **26.** $\cos A = 0.877$ **28.** $0.568, 1.22$ **30.** $2.002, 1.1178$
$\tan B = 1.83$

Section 13.3

2. 3.526 **4.** 10.08 **6.** 1.023 **8.** 1.247 **10.** 1.012 **12.** 1.265

14. $9°$ **16.** $66°$ **18.** $58°30'$ **20.** $30°20'$ **22.** $21°20'$ **24.** $66°30'$

26. 985 ft lb   **28.** $1,546$ ft lb   **30.** 0.9425 **32.** 1.1316 **34.** 0.7199

Section 13.4

2. $A = 54°$ **4.** $A = 50°40'$ **6.** $B = 46°$ **8.** $B = 75°40'$ **10.** $B = 77°50'$
$B = 36°$ $B = 39°20'$ $b = 873$ $a = 334$ $a = 17.6$
$c = 15.5$ $b = 104$ $c = 1214$ $b = 85$ $b = 3.8$

12. $B = 57°30'$ **14.** 184 m   **16.** 4.03 cm   **18.** 224 m   **20.** 4.4 cm, 10.4 cm
$a = 30.5$
$b = 47.9$

22. $13,600$ m   **24.** $30°40', 59°20', 16.5$ cm

Section 13.5

2. $-\dfrac{1}{2}, \dfrac{\sqrt{3}}{2}, -\dfrac{1}{\sqrt{3}}, -2, \dfrac{2}{\sqrt{3}}, -\sqrt{3}$ 4. $\dfrac{1}{2}, -\dfrac{\sqrt{3}}{2}, -\dfrac{1}{\sqrt{3}}, 2, -\dfrac{2}{\sqrt{3}}, -\sqrt{3}$

6. $\dfrac{1}{\sqrt{2}}, -\dfrac{1}{\sqrt{2}}, -1, \sqrt{2}, -\sqrt{2}, -1$ 8. $-\dfrac{1}{\sqrt{2}}, \dfrac{1}{\sqrt{2}}, -1, -\sqrt{2}, \sqrt{2}, -1$

10. -0.8910 12. $-.3256$ 14. -7.185 16. $-.9976$ 18. $.5225$

20. 3.021 22. $-.7813$ 24. $147°30'$ 26. $253°30'$ 28. $266°10'$

30. $174°$ 32. $167°10'$

Section 13.6

2. $(23, -25)$ 4. $(35, 19)$ 6. $(-7.2, 15.4)$ 8. $(4.80, 15.4)$ 10. $(-443, 232)$

12. $(25, 341°30')$ 14. $(47.5, 157°50')$ 16. $(3.84, 57°50')$ 18. $(76.2, 296°30')$

20. $(25.4, 14°20')$ 22. $(32.2, 240°20')$ 24. $V_x = 709$ cm/sec, $V_y = 472$ cm/sec

26. N $30°50'$ E   28. 363 kg   30. $33°40'$, 70 kg   32. 300 m/kg   34. 70 kg, 130 kg

36. AD = 163 and BD = 115; AD = 407 and BD = 190

Section 13.7

2. 2 4. $-7 + 6j$ 6. $-2 + j$ 8. $3 + 5j$ 10. $-3 + 5j$

12. $(2.9, 15), (2.09, -1.26)$ 14. $(-360, -130), (.48, .19)$

16. $(.019, -.118), (1.63, 2.52)$

Section 13.8

2. B = $22°$ 4. C = $39°20'$ 6. B = $14°$ 8. B = $20°40'$ 10. No triangle
    b = 66   b = 1.1   b = 11   C = $125°$
    a = 148   c = 1.9   c = 35   c = 10.4

12. A = $62°30'$ A' = $177°30'$ 14. 44 ft   16. 32 m   18. $117°$
    C = $74°$ C' = $19°$
    c = 5.3   c' = 1.8

20. 37.7 cm or 13.4 cm

Section 13.9

2. A = $13°$ 4. B = $53°10'$ 6. B = $61°40'$ 8. B = $108°50'$
    B = $128°$ C = $56°10'$ C = $47°30'$ A = $39°20'$
    c = 56   a = 61   a = 4.4   c = 36

10. A = $69°50'$ 12. B = $55°20'$ 14. $42°$ 16. 229 ft   18. 133 ft
    C = $82°10'$ C = $73°40'$
    b = 5.59   a = 68

20. $33.6°$

2.

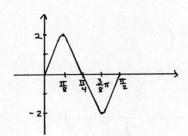

4.

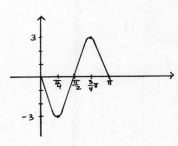

6.

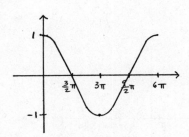

8.

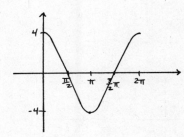

10.

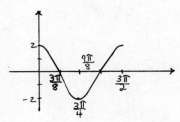

12.

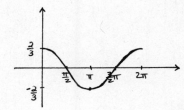

14.

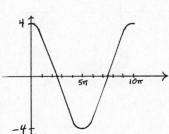

16.

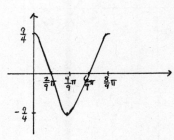

18.

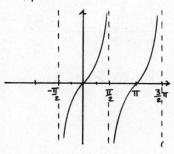

20.

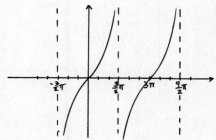

22.

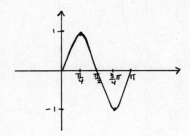

24.

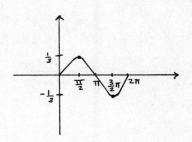

26.

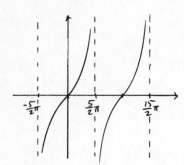

28.

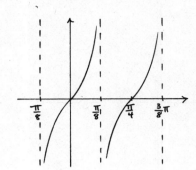

Section 13.11

2.

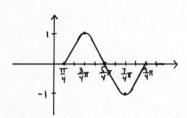

4.

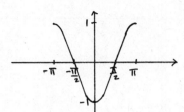

6.

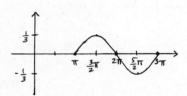

8.

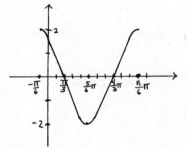

10.

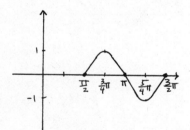

12.

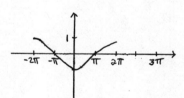

14.

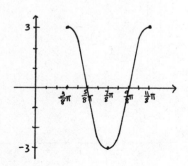

16.

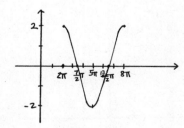

18.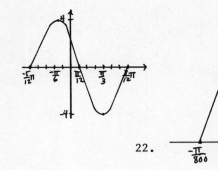

20.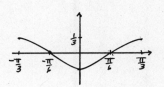

22.

Section 13.12

2. $x = 30°, 210°$ 4. $\theta = \frac{4\pi}{3}, \frac{5\pi}{3}$ 6. $\theta = \frac{5\pi}{6}, \frac{11\pi}{6}, \frac{\pi}{2}, \frac{3\pi}{2}$

8. $x = 90°, 270°$ 10. $x = 0°, 120°, 240°$ 12. $x = 113°30', 246°30', 109°30', 250°30'$

14. no solution   16. $x = 68°10', 248°10', 108°30', 288°30'$ 18. $x = 28°30', 208°30', 100°10', 280°10'$

20. $x = 34°40', 325°20'$ 22. $0.7297, 2.412, 3.553, 5.872$

51

ADDITIONAL TESTS

CHAPTER 1 PRETEST

1.1 1. 625
          3586
          92
        + 8564

2. 6000 – 1258 =

3. 97 – (19 + 43) =

4. (8 – 2) + 5(7 – 3) =

1.2 5. 2576
        x 504

6. 27m x 38m =

7. 756 ÷ 462 =

8. 50 kl ÷ 25 min =

9. 72 ÷ 8 + (37 – 14) =

1.3 10. Write using exponents: 4 x 4 x 4 x 4

11. 2^3 x 3^2 =

1.4 12. List the prime numbers that are greater than 10 and less than 25.

13. Find the prime factorization of 400.

1.5 14. Express $\frac{32}{17}$ as a mixed number.

15. Express $13\frac{2}{5}$ as an improper fraction.

1.6 16. $\frac{4}{5}$ = $\frac{}{30}$

17. $\frac{13}{24}$ = $\frac{}{72}$

1.7 18. Reduce $\frac{48}{54}$ to lowest terms using prime factorization.

19. Reduce $\frac{125}{175}$ to lowest terms using prime factorization.

1.8 20. Find the lowest common denominator of $\frac{3}{8}$ and $\frac{5}{6}$ and then express them as equivalent fractions with the lowest common denominator.

1.9 21. $\frac{7}{15}$ – $\frac{4}{15}$ =

22. $\frac{1}{3}$ + $\frac{1}{4}$ + $\frac{1}{5}$ =

23. $4\frac{1}{5}$ + $3\frac{1}{4}$ =

24. $7\frac{4}{7}$ – $3\frac{5}{7}$ =

25. $6\frac{2}{3}$ – $2\frac{3}{4}$ =

1.10 26. $\frac{5}{6}$ x $\frac{18}{35}$ =

27. $5\frac{3}{8}$ x $2\frac{1}{6}$ =

28. $\frac{9}{15}$ ÷ $\frac{3}{5}$ =

29. $8\frac{1}{4}$ ÷ $2\frac{1}{2}$ =

1.11   30.   Write in words: 1507.

        31.   Write in decimal notation: eight and fifty-six thousandths.

        32.   How many significant digits does 0.0050 contain?

        For problems 33 through 36, perform the indicated operations assuming the given data are approximations:

1.12   33.   15.6 - 2.78 =   34.   2.63 + 0.0042 + 15.06 =

1.13   35.   6.3 x 5000 =   36.   1.75 $\div$ 0.2 =

        For problems 37 and 38, perform the indicated operations on the calculator:

1.14   37. 12^9 38. $\sqrt{189}$

1.15   39.   Express 39.13 as a fraction.

        40.   Express $\frac{3}{8}$ as a decimal.

1.16   41.   Express 0.026 as a percent.

        42.   Express $\frac{5}{8}$ as a percent.

        43.   Express 63.5% as a decimal.

        44.   Express 55% as a fraction.

1.17   45.   Change 6.8 m to centimeters.

For problems 1 through 13, perform the indicated operations:

1. 732 + 6568 + 68 + 724   2. 3704 − 497

3. 6257
    x 708   4. 35,872 ÷ 59

5. 52 cm ÷ 4 cm   6. 7.92 − 0.773

7. 6.84 + 9.2 + 13.79   8. 0.4094 ÷ 0.270

9. 6.41 x 0.55   10. $\frac{3}{8} + \frac{5}{6} + \frac{1}{12}$

11. $\frac{3}{5} - \frac{1}{3}$ 12. $6\frac{3}{7} - 2\frac{5}{7}$

13. $33 ÷ 5\frac{1}{2}$

14. Indicate the number of significant digits in each of the following:

    a. 62,004   b. 31.25   c. 0.040   d. 6.020

15. Round off each of the following to the indicated place:

    a. 3.55 (tenths)   b. 16,894.7 (thousandths)

    c. 7.8694 (hundredths)   d. 9.8 (units)

16. Give the prime factorization of 1032.

17. Reduce: $\frac{1155}{2695}$

18. Write with exponents: 6 x 6 x 6 x 6

19. Express in expanded notation using powers of ten: 4285

20. Express as an equivalent number in scientific notation: 2876

21. Change to percent: $\frac{2}{5}$

1. _____
2. _____
3. _____
4. _____
5. _____
6. _____
7. _____
8. _____
9. _____
10. _____
11. _____
12. _____
13. _____
14.a._____ b._____
   c._____ d._____
15.a._____ b._____
   c._____ d._____
16. _____
17. _____
18. _____
19. _____
20. _____
21. _____

22. Change to a fraction in lowest form: 36%   22._____

23. Change to the indicated measure:

    a. 45 centigrams to hectograms   23a._____

    b. 1.65 liters to centiliters   23b._____

24. Change to grams: 74 pounds   24._____

25. 168 + 78 x 23 − 344 ÷ 43   25._____

CHAPTER 1 TEST
FORM B

For problems 1 through 13, perform the indicated operations:

1. 6407 + 98 + 613   2. 4003 - 309   1._____

3. 98 x 675   4. 14,652 ÷ 36   2._____

                                                              3._____

5. 16 m x 58   6. 62.14 - 0.866   4._____

                                                              5._____

7. 24.70 + 3.18 + 0.047   8. 0.01242 ÷ 0.270   6._____

9. 7.04 x 0.26   10. $\frac{7}{10} + \frac{3}{15} + \frac{9}{20}$ 7._____

                                                              8._____

11. $\frac{7}{11} - \frac{1}{3}$ 12. $6\frac{1}{5} - 2\frac{1}{2}$ 9._____

                                                              10._____

13. $\frac{6}{7} \div 3\frac{1}{5}$ 11._____

                                                              12._____

14. Indicate the number of significant digits in each of the
     following: 13._____

     a. 7030   b. 6.4   c. 0.0060   d. 71.050   14.a._____b._____

                                                                 c._____d._____
15. Round off each of the following to the indicated place:
                                                              15.a._____b._____
     a. 3.85 (tenths)   b. 178.4 (hundredths)
                                                                 c._____d._____
     c. 16.006 (hundredths)   d. 784.78456 (ten thousandths)

16. Find the prime factorization: 3690   16._____

17. Reduce: $\frac{91,728}{51,952}$ 17._____

18. Write without exponents and simplify: 5^3 18._____

19. Express in expanded notation using powers of ten: 659   19._____

20. Express as an equivalent number in scientific notation: 972   20._____

21. Change to percent: $\frac{3}{10}$ 21._____

57

22. Change to a fraction in lowest form: 64%

23. Change to the indicated measure:

 a. 684 km to m   b. 16.82 ml to cl

24. Change 3 feet to meters.

25. 4678 − 13 x 27 + 1674 ÷ 9

22._____

23._____

 a._____ b._____

24._____

25._____

CHAPTER 1 TEST
FORM C

For problems 1 through 13, perform the indicated operations:

1. 618 + 407 + 398 + 4731

2. 3004 − 1729

3. 4653
    x 839

4. 251,082 ÷ 39

5. 8 m x 7

6. 164 cm ÷ 4 cm

7. 63.2 − 0.774

8. 57.6 x 4.06

9. 190.037 − 3.07

10. $\frac{4}{9} + \frac{2}{3} + \frac{5}{6}$

11. $\frac{4}{15} - \frac{1}{10}$

12. $7\frac{1}{3} - 2\frac{2}{3}$

13. $9 \div 1\frac{1}{5}$

14. Indicate the number of significant digits in each of the
     following:

     a. 8.500   b. 0.0365   c. 2070   d. 28.62

15. Round off each of the following to the indicated place:

     a. 6.0578 (thousandths)   b. 6.5 (units)

     c. 684.26 (thousandths)   d. 840.06 (hundredths)

16. Give the prime factorization: 12,375

17. Change to decimal form: $\frac{5}{11}$

18. Change to a fraction in lowest form: 0.058

19. Change to percent: 0.045

20. Change to a fraction: 6.8%

21. Change 446.4 cm^2 to mm^2

22. Change 468 mg to kg

1. _____

2. _____

3. _____

4. _____

5. _____

6. _____

7. _____

8. _____

9. _____

10. _____

11. _____

12. _____

13. _____

14.a. _____ b. _____

c. _____ d. _____

15.a. _____ b. _____

16. _____

17. _____

18. _____

19. _____

20. _____

21. _____

22. _____

23. Change 18,420 m^3 to dam^3

23._____

24. Change 6.2 pounds to grams

24._____

25. How many pieces of pipe, each measuring 6.5 cm, can be cut from a piece of pipe measuring 2.6 m?

25._____

For problems 1 through 13, perform the indicated operations:

1. $604 + 7349 + 77 + 835$ 2. $2603 - 386$

3. 5073×608 4. $50,468 \div 74$

5. $18 \text{ m} \times 37 \text{ m}$ 6. $3717 \text{ cm} \div 59 \text{ cm}$

7. $6.80 - 0.662$ 8. 6.42×0.107

9. $75.67 \div 4.10$ 10. $\frac{2}{7} + \frac{3}{14} + \frac{5}{4}$

11. $\frac{5}{7} - \frac{1}{8}$ 12. $5\frac{3}{5} - 2\frac{4}{5}$

13. $14 \div 3\frac{1}{2}$

14. Indicate the number of significant digits in each of the following:

   a. 7030   b. 6.4   c. 7.200   d. 0.00400

15. Round off each of the following to the indicated place:

   a. 3.5 (units)   b. 6843.5 (thousandths)

   c. 60.625 (hundredths)   d. 4.358 (tenths)

16. Find the prime factorization: 8232

17. Change to decimal form: $\frac{3}{7}$

18. Change to a fraction in lowest form: 0.066

19. Change to a percent: 0.43

20. Change to a fraction: 1.75%

21. Change 14.52 dam^3 to m^3

22. Change 468,000 mg to kg

1._____

2._____

3._____

4._____

5._____

6._____

7._____

8._____

9._____

10._____

11._____

12._____

13._____

14.a._____ b._____

   c._____ d._____

15.a._____ b._____

   c._____ d._____

16._____

17._____

18._____

19._____

20._____

21._____

22._____

23.   Change 2 ounces to milligrams   23._____

24.   Change 681 ml to l   24._____

25.   How much change will John receive from a twenty-dollar   25._____
       bill if he purchases three bags of concrete mix at $1.89
       a bag and five cans of spray paint at $2.07 a can?

ANSWERS FOR CHAPTER 1 PRETEST

1. 12,867

2. 4742

3. 35

4. 25

5. 1,298,304

6. 1026 m^2

7. 1 R 294

8. 2 kl/min

9. 32

10. 4^4

11. 72

12. 11, 13, 17, 19, 23

13. $2^4 \times 5^2$

14. $1\frac{15}{17}$

15. $\frac{67}{5}$

16. 24

17. 39

18. $\frac{8}{9}$

19. $\frac{5}{7}$

20. LCD = 24; $\frac{9}{24}$; $\frac{20}{24}$

21. $\frac{11}{15}$

22. $\frac{47}{60}$

23. $7\frac{9}{20}$

24. $3\frac{6}{7}$

25. $3\frac{11}{12}$

26. $\frac{3}{7}$

27. $11\frac{31}{48}$

28. 1

29. $1\frac{3}{10}$

30. one thousand five hundred seven

31. 8.056

32. 2

33. 12.8

34. 18

35. 30,000

36. 9

37. 5,159,780,351

38. 13.75

39. $\frac{3913}{100}$

40. 0.375

41. 2.6%

42. 62.5%

43. 0.635

44. $\frac{11}{20}$

45. 680 cm

ANSWERS TO CHAPTER 1 TESTS

FORM A

1. 8092
2. 3207
3. 4,429,956
4. 608
5. 13
6. 7.15
7. 29.8
8. 1.52
9. 3.5
10. $\frac{31}{24}$ or $1\frac{7}{24}$
11. $\frac{4}{15}$
12. $3\frac{5}{7}$
13. 6
14a. 5
14b. 4
14c. 2
14d. 4
15a. 3.6
15b. 17,000
15c. 7.87
15d. 10
16. $2^3 \cdot 3 \cdot 43$
17. $\frac{3}{7}$
18. 6^4
19. $(4 \times 10^3) + (2 \times 10^2) + (8 \times 10^1) + (5 \times 10^0)$
20. 2.876×10^3

21. 40%
22. $\frac{9}{25}$
23a. 0.0045 hg
23b. 165 cl
24. 33,565.66 g
25. 1954

FORM B

1. 7118
2. 3694
3. 66,150
4. 407
5. 928 m
6. 61.27
7. 27.93
8. 0.046
9. 1.83
10. $\frac{27}{20}$ or $1\frac{7}{20}$
11. $\frac{10}{33}$
12. $3\frac{7}{10}$
13. $\frac{15}{56}$
14a. 3
14b. 2
14c. 2
14d. 5
15a. 3.9
15b. 200
15c. 16.01
15d. 784.7846
16. $2 \cdot 3^2 \cdot 5 \cdot 41$
17. $\frac{13}{17}$
18. 125
19. $(6 \times 10^2) + (5 \times 10^1) + (9 \times 10^0)$
20. 9.72×10^2

21. 30%
22. $\frac{16}{25}$
23a. 684,000 m
23b. 1.682 cl
24. 0.9144 m
25. 4513

ANSWERS TO CHAPTER 1 TEST (continued)

FORM C

1. 6154   21. 44,640 mm^2

2. 1275   22. 0.000468 kg

3. 3,903,867   23. 18.42 dam^3

4. 6438   24. 2800 g

5. 56 m   25. 40

6. 41

7. 62.4

8. 234

9. 61.9

10. $\frac{35}{18}$ or $1\frac{17}{18}$

11. $\frac{1}{6}$

12. $4\frac{2}{3}$

13. $7\frac{1}{2}$

14a. 4

14b. 3

14c. 3

14d. 4

15a. 6.058

15b. 7

15c. 1000

15d. 800

16. $3^2 \cdot 5^3 \cdot 11$

17. 0.454545...

18. $\frac{29}{500}$

19. 4.5%

20. $\frac{68}{1000}$ or $\frac{17}{250}$

FORM D

1. 8865   21. 14520 m^3

2. 2217   22. 0.468 kg

3. 3,084,384   23. 60,000 mg

4. 682   24. 0.681 l

5. 666 m^2 25. $3.98

6. 63

7. 6.14

8. 0.687

9. 18.5

10. $\frac{7}{4}$ or $1\frac{3}{4}$

11. $\frac{33}{56}$

12. $2\frac{4}{5}$

13. 4

14a. 3

14b. 2

14c. 4

14d. 3

15a. 4

15b. 7000

15c. 60.62

15d. 4.4

16. $2^3 \cdot 3 \cdot 7^3$

17. 0.428571428571...

18. $\frac{33}{500}$

19. 43%

20. $\frac{7}{400}$

2.1   For problems 1 through 3, consider the following figure:

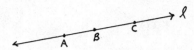

1. Name $\overrightarrow{CB}$ in another way.
2. Name two rays on line ℓ with endpoint B.
3. Name three line segments on line ℓ.

2.2   4. Find the complement of an angle whose measure is 62°19' 48".
       5. Find the supplement of an angle whose measure is 113°26' 50".

2.3   6. Find the perimeter of the following triangle ABC:

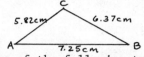

       7. Find the area of the following triangle ABC:

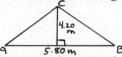

2.4   8. Find the perimeter and area of the following polygon:

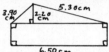

2.5   9. Find the circumference of a circle whose radius is 5.00 cm.
       10. Find the area of a circle whose diameter is 4.86 m.

2.6   11. Find the surface area of a cube one of whose edges is 5 cm.
       12. Find the volume of a sphere whose diameter is 6.4 m.

CHAPTER 2 TEST
FORM A

Given the line b, and points M, N, and P on the line:

1. Name three different line segments on b.
2. How many end-points does b have?
3. Name four different rays on b.
4. Name a line coincident with b.

Given: r ∥ s

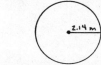

5. ∠FEH and ∠_____ are vertical angles.
6. ∠BED and ∠_____ are alternate interior angles.
7. _____ is a straight angle.
8. _____ is an acute angle.

Find the perimeter and the area of the following:

9.
3.2m 2m 5.9m
   6.4m

10.
3 in.   5 in.
   8 in.

11.
14.1cm
5cm 14.1 cm 5cm
   10.1cm

12.
7.4cm 1.4 cm 2.5cm
7.3cm   2.1cm
7.4cm 1.4 cm 2.5cm

13. Find the circumference and area of the following:

2.14 m

14. Find the area of the following:

4.3cm
   4.3 cm

15. Find the complement of an angle whose measure is 37°.

16. Find the supplement of an angle whose measure is 62°.

17. Is an angle whose measure is 91° a right angle or an obtuse angle?

18. Find the surface area of a cube if one of its edges is 5 ft.

Find the surface area and volume of the following:

19.
22.62 cm
8.64 cm

20.

12.7 cm   8.4cm
   11.6cm

1._____
2._____
3._____
4._____
5._____
6._____
7._____
8._____
9._____
10._____
11._____
12._____
13._____
14._____
15._____
16._____
17._____
18._____
19._____
20._____

Given the line p and point B, C, and E on the line:

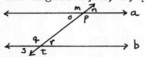

1. Name a line coincident with p.
2. Name three different rays on p.
3. Name three different line segments on p.
4. How many end-points does $\overrightarrow{BE}$ have?

Given: a ∥ b and angles m, n, o, p, q, r, s and t:

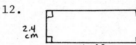

5. Name four interior angles.
6. Name four exterior angles.
7. ∠q and ∠____ are alternate interior angles.
8. ∠s and ∠____ are alternate exterior angles.
9. ∠n and ∠____ are corresponding angles.
10. ∠m and ∠____ are vertical angles.

Find the perimeter and area of the following:

1._____
2._____
3._____
4._____
5._____
6._____
7._____
8._____
9._____
10._____
11._____
12._____
13._____
14._____
15._____
16._____

11.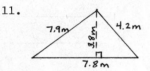

12.

13.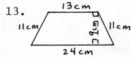

14.

15.

16. Find the circumference and area of a circle with a diameter of 12.5 m.

17. Find the area of the shaded area:

17._____

18. Find the complement of an angle whose measure is 71°.

18._____

19. Find the supplement of an angle whose measure is 86°.

19._____

20. Is an angle whose measure is 179° an obtuse angle or a straight angle?

20._____

21. If the measures of two angles of a triangle are 36° and 73°, what is the measure of the third angle?

21._____

22. Find the surface area and volume of a cube whose edge is 4 cm.

22._____

23. Find the volume of a regular prism whose base is a triangle of 30 cm² and whose altitude is 8 cm.

23._____

24. Find the surface area of a regular prism whose base is a rectangle 7 cm by 9 cm and whose altitude is 5 cm.

24._____

25. Find the surface area and volume of a sphere with diameter 22.4 cm.

25._____

Given the line p, points H and K on the line:

1. Name a line coincident with p.
2. How many end-points does $\overrightarrow{HK}$ have?
3. Name a line segment on p.
4. Name two rays on p.

For problems 5 through 8, find the perimeter and area:

5.

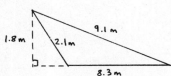

6.

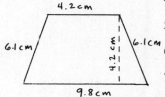

7.

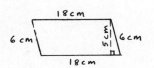

8.

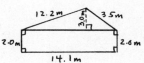

Given: 1 ∥ m

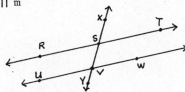

9. ∠RSV and _____ are alternate interior angles.
10. _____ is an obtuse angle.
11. _____ is an acute angle.
12. ∠RSX and _____ are vertical angles.

Find the circumference and area of the circles, given:

13. radius = 3.6 cm   14. diameter = 6.4 m

15. Find the area of a circular ring with inside diameter of 5 cm
    and an outside diameter of 7 cm.

16. Find the area of a cube whose edge is 9 cm.

17. Find the surface area and volume of a regular circular cylinder
    whose base has a diameter of 4.01 cm and whose altitude is
    21.2 cm.

18. Find the surface area and volume of a regular prism with an
    8 m by 10 m rectangular base and 12 m altitude.

19. Find the surface area and volume of a regular circular cone with
    a base of radius 9.00 cm, altitude of 22.0 cm, and whose lateral
    height is 22.8 cm.

1._____
2._____
3._____
4._____
5._____
6._____
7._____
8._____
9._____
10._____

9._____
10._____
11._____
12._____
13._____
14._____
15._____
16._____
17._____
18._____
19._____

20. Find the surface area and volume of a sphere with radius 4.4 m.

20._____

CHAPTER 2 TEST
FORM D

Given the line ℓ and points E, F, and K on the line:

1. Name a line coincident with ℓ.
2. How many end-points does $\overline{EK}$ have?
3. Name three different rays on ℓ.
4. Name two different line segments on ℓ.

Given: AC || MN

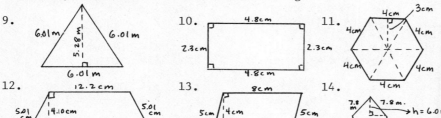

5. Name a line which intersects $\overleftrightarrow{AC}$.
6. Name two alternate exterior angles.
7. Name one straight angle.
8. Name a pair of vertical angles.

Find the perimeter and area of the following:

9. triangle: 6.01 m, 6.01 m, 5.28 m (height), 6.01 m (base)

10. rectangle: 4.8cm, 2.3cm, 2.3cm, 4.8cm

11. hexagon: 3cm, 4cm, 4cm, 4cm, 4cm, 4cm, 4cm

12. trapezoid: 12.2 cm, 5.01 cm, 4.10 cm, 5.01 cm, 14.1 cm

13. parallelogram: 8cm, 5cm, 4cm, 5cm, 8cm

14. 7.8 m, 7.8 m, h=6.0 m, 8.0 m, 8.0 m, 10 m

15. Find the circumference and area of a circle with radius 9.6 m.

16. Find the circumference and area of a circle with diameter 6.8 cm.

17. Find the area of a washer 24.08 mm in diameter with a hole whose diameter is 6.66 mm.

18. Find the surface area and volume of a cube whose edge is 5.24 cm.

19. Find the surface area and volume of a regular circular cylinder whose base has a diameter of 16.2 dm and whose altitude is 27.8 dm.

20. Find the surface area and volume of a sphere 31.6 cm in radius.

1. _____
2. _____
3. _____
4. _____
5. _____
6. _____
7. _____
8. _____
9. _____
10. _____
11. _____
12. _____
13. _____
14. _____
15. _____
16. _____
17. _____
18. _____
19. _____
20. _____

71

ANSWERS TO CHAPTER 2 TESTS

PRETEST	FORM A	FORM B
1. $\overrightarrow{CA}$	1. $\overline{MN}$, $\overline{MP}$, $\overline{NP}$	1. $\overleftrightarrow{BC}$ or $\overleftrightarrow{BE}$ or $\overleftrightarrow{CE}$
2. $\overrightarrow{BA}$ and $\overrightarrow{BC}$	2. 0	2. $\overrightarrow{BC}$ or $\overrightarrow{BE}$, $\overrightarrow{CE}$, $\overrightarrow{EC}$ or $\overrightarrow{EB}$
3. $\overline{AB}$, $\overline{BC}$, $\overline{AC}$	3. $\overrightarrow{MN}$ or $\overrightarrow{MP}$, $\overrightarrow{NP}$, $\overrightarrow{PN}$ or $\overrightarrow{PM}$, $\overrightarrow{NM}$	3. $\overline{BC}$, $\overline{BE}$, $\overline{CE}$
4. 27°40'12"	4. $\overleftrightarrow{MN}$ or $\overleftrightarrow{MP}$ or $\overleftrightarrow{NP}$	4. 1
5. 66°33' 10"	5. ∠BED	5. ∠o, ∠p, ∠s, ∠r
6. 19.4 cm	6. ∠EDG	6. ∠m, ∠n, ∠s, ∠t
7. 12.2 m^2	7. ∠ADG	7. p
8. Perimeter: 23.3 cm	8. ∠BED or ∠EDG	8. n
Area: 31.9 cm^2	9a. P = 15.5 m	9. r
9. 314 cm	9b. A = 6.4 m^2	10. p
10. 18.5 m^2	10a. P = 26 in.	11a. A = 15 m^2
11. 150 cm^2	10b. A = 24 in.^{2}	11b. P = 19.9 m
12. 113 m^3	11a. P = 34.2 cm	12a. A = 9.8 cm^2
	11b. A = 49.61 cm^2	12b. P = 12.9 cm
	12a. P = 19.8 cm	13a. A ≐ 200 cm^2
	12b. A = 13.16 cm^2	13b. P = 59 cm
	13a. C = 13.4 m	14a. A ≐ 11.1 cm^3
	13b. A = 14.4 m^2	14b. P = 14.5 cm
	14. 14.5 cm^2	15a. A = 42 in.^{2}
	15. 53°	15b. P = 26 in.
	16. 118°	16. 39.3 m, 122.7 m^2
	17. obtuse angle	17. 414.48 m^2
	18. 150 ft^2	18. 19°
	19a. SA = 730.9 cm^2	19. 94°
	19b. V ≐ 1330 cm^3	20. obtuse angle
	20a. SA ≐ 890 cm^2	21. 71°
	20b. V ≐ 1200 cm^3	22. 96 cm^2, 64 cm^3
		23. 240 cm^3
		24. 286 cm^2
		25. SA = 1,576 cm^2
		V = 5,882 cm^3

ANSWERS TO CHAPTER 2 TESTS (continued)

FORM C

1. $\overleftrightarrow{HK}$

2. 1

3. $\overline{HK}$

4. $\overrightarrow{HK}$, $\overrightarrow{KH}$

5. P = 19.5 m
   A = 7.5 m^2

6. P = 26.2 cm
   A = 29 cm^2

7. P = 48 cm
   A = 90 cm^2

8. P = 33.8 m
   A = 49 m^2

9. $\angle$ SVW

10. $\angle$ RSX or $\angle$ UVS

11. $\angle$ XST or $\angle$ SVW

12. $\angle$ TSV

13. 22.61 cm, 40.69 cm^2

14. 20.10 m, 32.15 m^2

15. 75.4 cm^2

16. 486 cm^2

17. 293 cm^2, 268 cm^3

18. 592 m^2, 960 m^3

19. 899 cm^2, 1860 cm^3

20. 243 m^2, 357 m^3

FORM D

1. $\overleftrightarrow{EF}$ or $\overleftrightarrow{EK}$ or $\overrightarrow{FK}$

2. 2

3. $\overrightarrow{FE}$, $\overrightarrow{KE}$, $\overrightarrow{EF}$, $\overrightarrow{EK}$, $\overrightarrow{FK}$

4. $\overline{KF}$, $\overline{FE}$, $\overline{KE}$

5. BE

6. $\angle$HBC & $\angle$DEG; $\angle$ABH & $\angle$GEF

7. $\angle$ABC or $\angle$DEF

8. $\angle$CBH & $\angle$ABE; $\angle$ABH & $\angle$EBC;
   $\angle$BEF & $\angle$DEG; $\angle$DEB & $\angle$GEF

9. P = 18.0 m
   A = 15.9 m^2

10. P = 14 cm
    A ≐ 11 cm^2

11. P = 24 cm
    A = 36 cm^2

12. P = 36.3 cm
    A = 53.9 cm^2

13. P = 26 cm
    A = 32 cm^2

14. P ≐ 42 m
    A = 110 m^2

15. 60 m, 289 m^2

16. 21 cm, 36 cm^2

17. 420.4 mm^2

18. 165 cm^2, 144 cm^3

19. 1,830 dm^2, 5,730 dm^3

20. 12,500 cm^2
    132,000 cm^3

CHAPTER 3 PRETEST

3.1   1.   Determine which real number is the largest:

              -18   -14   -22

       2.   Give the negative of -13.

       3.   Evaluate $\left|-46\right|$.

       4.   Insert the sign $>$ or $<$ between the two real numbers: -11   -4

In problems 5 through 11, perform the indicated operations.

3.2   5.   -4 + (-18)   6.   -6 + 8   7. -5 + 9 + (-3)   8. -1 + (-4) + 6

3.3   9.   9 - 15   10.   -8 - (-11)   11. -2 + 5 - 3 + 7 - 1 - (-9)

3.4   In problems 12 through 16, add or subtract the following:

     12.   16Y + 3Y   13.   18R - (-9R)   14.   e + 3e - (-2e)

     15.   (5M + 3) + (2M - 6)   16.   (2a + 3b - c) - (a + b - c)

3.5   17.   Multiply: -2(-13)

     18.   Divide: (-44) ÷ 11

3.6   19.   Multiply: $-2a^4(3a^5)$

     20.   Perform the indicated operations:

                 3(2N + 4) - 2(3N - 5)

3.7   21.   Divide: $9Y^8 ÷ (-3Y^4)$

     22.   Divide: $(80e^5 + 90e^4 - 20e) ÷ 10e^2$

3.8   Remove all symbols of grouping and simplify:

     23.   6A - [7 - (10A - 3) + 7]

     24.   6(4 + Y) - [3(2Y - 3) + 3Y] - 11Y

3.9   25.   Evaluate the given algebraic expression, given a = 2, b = -3, and c = -1.

$$a^2 - abc - c^2$$

Evaluate the following:

1. $-10 - 4$ 2. $(-3) + (-1) + (-4)$

3. $(-4) - (-3)$ 4. $16 - (-20)$

5. $(6)(24) + (-8.4)$ 6. $144 \div (-6)$

7. $(-3/4)(-12/7)$ 8. $4rs^2(-20r^2s^3)$

9. $-3ab^2(5a - 2b)$

10. $(4L_0 - L_1 + 2L_2) + (-3L_0 - L_1 - L_2) + (5L_0 - 2L_1 - 3L_2)$

11. $(5x - 2y - 6) - (2x - 3y - 1)$

12. $(-2A + B - 4) - (-2A - B + 6)$

13. $28U^3V^2 \div -14UV$

14. $\dfrac{15C^4 - 10C^2 - 5C}{-5C}$ 15. $\dfrac{63p + 7p^2 - 56p^3}{-7p}$

16. $4m - 2(m - 5) + 6$ 17. $16 - [2T - (3T - 1) + 6]$

18. If $a = -1$, $b = 2$, and $c = 0$, find the value of:

$$2(3a - b) + 6c$$

19. If $m = 4$, $n = -2$, and $p = -3$, find the value of:

$$mn + n - (p + 6)$$

20. Find S, given $S = \dfrac{3L(W - L)}{8}$

     $L = 15.5$ ft
     $W = 16$ ft

1._____

2._____

3._____

4._____

5._____

6._____

7._____

8._____

9._____

10._____

11._____

12._____

13._____

14._____

15._____

16._____

17._____

18._____

19._____

20._____

CHAPTER 3 TEST
FORM B

1. Give the negative of –11.

2. Evaluate $\left| -\dfrac{3}{7} \right|$

3. Insert $<$ or $>$ between the given numbers:

   a. –3   –11   b. 6 – 8   9 – 1

For problems 4 through 31, perform the indicated operations:

4. –6 – 8   5. – 6 – (–9)   6. 4 – 7

7. 6 + (–10)   8. –11 + (–2)   9. 3/10 + (–4/15)

10. 6 + 8 – 3 – 7 + 4

11. –3 – (–5) + (–7) + 8 – 9   12. 6(–7)

13. (–3)(–5)   14. (–6)(–1)(–8)   15. (–5/7)(21/40)

16. (–63)(–7)   17. (2/5) ÷ (–5/3)

18. $b^5 \cdot b^3$ 19. $(-5a^4)(-2b)$

20. $y^7 \div y^2$ 21. $(-32x^5) \div (2x)$

22. $(3cd^2)(-4cd)(-2c^2d)$ 23. 8 + (–10)(2) – 16 ÷ (–2)

24. (2A – 5B + 6) + (–3A – 4B – 8)

25. (7c – 2d – 5) – (4c – 3d – 5)

26. Subtract $(3R_0 – 3R + 3)$ from the sum of $(2R_0 + R + 1)$ and $(-3R_0 – R + 5)$.

27. $-2x^2y(3xy – 2xy^2 – 5)$ 28. $\dfrac{28e^4f^7 – 36e^2f^4 + 20e^2f^2}{-4e^2f^2}$

29. 6 + [3m – (2m – 3) + 7]

30. 2T + 4 – [4T – (2T – 6)]

31. $2a – \left\{ 6a – [4(a – 3) – 6] + 1 \right\}$

32. Find the value of $a^2 – b^2 + c$ if a = –3, b = –2, and c = –1.

33. Find the value of 2xy(2x – 3y – 4) if x = –4 and y = –3.

34. Given: $L = 2D + \dfrac{13}{14} (R + r)$. Find L when:

   D = 4 ft, R = 20 in., and r = 4 in.

1._____

2._____

3a._____ b._____

4._____

5._____

6._____

7._____

8._____

9._____

10._____

11._____

12._____

13._____

14._____

15._____

16._____

17._____

18._____

19._____

20._____

21._____

22._____

23._____

24._____

25._____

26._____

27._____

28._____

29._____

30._____

31._____

32._____

33._____

34._____

CHAPTER 3 TEST
FORM C

1. Give the negative of −16.   1._____

2. Evaluate: $\left|-16\right|$.   2._____

3. Insert $<$ or $>$ between the given: 3a._____

   a. −8   −1   b. −2 + 6   3 − 4   3b._____

4. −5   −6   4._____

5. −4 + (−5) − (−2)   5._____

6. 2(−8) − (−10.4)   6._____

7. 128 ÷ (−8)   7._____

8. (−2/7)(−5/8)   8._____

                                                                 9._____

9. −20 ÷ 2 − 16(−4)

10. (2R − 3S + 5) + (5R + 7S − 2)   10._____

11. $(-3T_0 + 4T_1 - 6T_2) - (2T_0 - 2T_1 + T_2)$ 11._____

12. $5xy^2(-14x^2y^3)$ 12._____

13. $-2ab^2(6a - 2b)$ 13._____

14. $-38m^6v^3 \div 19m^2v^4$ 14._____

15. $\dfrac{15d^6 - 12d^4 - 3d^2}{-3d^2}$ 15._____

                                                                16._____

16. H − [2 − 3(H − 4) + 4H]   17._____

17. $2j - 3\left\{j - [4 - 2j + 6) + j] - 3\right\}$

18. Find the value of 3x − 2y if: 18._____

    x = −3, y = 2

19. Find the value of $a^2 - b^2 - c + bc$ if: 19._____

    a = −1, b = 4, and c = −2

20. Find s, given the formula: 20._____

    $s = \dfrac{3L(W - L)}{9}$, and L = 21cm, W = 42 cm.

CHAPTER 3 TEST
FORM D

1. Give the negative of –20.

2. Evaluate $\left| -7 \right|$

3. Insert $<$ or $>$ between the given numbers:

   a. –2   –7   b. –6   4

For problems 4 through 33, perform the indicated operations:

4. –2 – 12   5. –6 – (–5)   6. 8 – 10

7. 4 + (–9)   8. –20 + (–6)   9. 2/7 + (–3/14)

10. 8 + 4 – 6 – 2 + 10   11. –2 – (–5) + (–6) + 9 – 1

12. 5(–8)   13. –6(–4)   14. –2(–1)(–11)

15. –3/4(12/27)   16. 3/9 ÷ (–3/10)   17. $a^6 \cdot a^4$

18. $-2b^2(-3b)$ 19. $x^8 \div x^6$ 20. $-20y^4 \div (-4y)$

21. $2ab^2(-3a^2b)(-2ab)$ 22. 6 + (–3)(–2) – 14 ÷ 7

23. (4H – 3K + 7) + (–2H – 4K + 6)

24. (12f – 3g – 6) – (2f – 7g – 2)

25. Subtract $(2M_0 - 6M + 4)$ from the sum of:

     $(5M_0 - 2M - 3)$ and $(-2M_0 + M + 5)$

26. $-4d^2c(2dc - 3dc^2 - 8)$

27. $(21R^3S^5 - 28R^4S^3 + 42R^2S) \div -7R^3S$

28. 5 + [6a – (3a – 4) + 1]

29. 2T + 8 – [5T – (3T + 7)]

30. $3b - \left\{ 5b - [5(b-4) - 9] + 6 \right\}$

31. Find the value of $a^2 + b^2 - c$ if:

     a = –2, b = 2, and c = –1.

32. Find the value of 3xy(3x – 4y + 8) if:

     x = –3, and y = –2.

33. Given that $L = 2D + \dfrac{13}{4}(R + r)$, find L when:

     D = 6 ft, R = 16 in., and r = 4 in.

1. _____

2. _____

3a. _____ b. _____

4. _____

5. _____

6. _____

7. _____

8. _____

9. _____

10. _____

11. _____

12. _____

13. _____

14. _____

15. _____

16. _____

17. _____

18. _____

19. _____

20. _____

21. _____

22. _____

23. _____

24. _____

25. _____

26. _____

27. _____

28. _____

29. _____

30. _____

31. _____

32. _____

33. _____

ANSWERS FOR CHAPTER 3 TESTS

PRETEST

1. -14
2. 13
3. 46
4. $<$
5. -22
6. 2
7. 1
8. 1
9. -6
10. 3
11. 15
12. 19Y
13. 27R
14. 6e
15. 7M - 3
16. a + 2b
17. 26
18. -4
19. $-6a^9$
20. 22
21. $-3Y^4$
22. $8e^3 + 9e^2 - \dfrac{2}{e}$
23. 16A - 17
24. 33 - 14Y
25. -3

FORM A

1. -14
2. -8
3. -1
4. 36
5. 135.6
6. -24
7. $\dfrac{9}{7}$
8. $-80r^3s^5$
9. $-15a^2b^2 + 6ab^3$
10. $6L_0 - 4L_1 - 2L_2$
11. 3x + y - 5
12. 2B - 10
13. $-2U^2V$
14. $-3C^3 + 2C + 1$
15. $-9 - p + 8p^2$
16. 2m + 16
17. T + 9
18. -10
19. -13
20. S = 2.90625

FORM B

1. 11
2. $\dfrac{3}{7}$
3a. $>$
 b. $<$
4. -14
5. 3
6. -3
7. -4
8. -13
9. $\dfrac{1}{30}$
10. 8
11. -6
12. -42
13. 15
14. -48
15. $-\dfrac{3}{8}$
16. 441
17. $-\dfrac{6}{25}$
18. b^8
19. $10a^4b$
20. y^5
21. $-16x^4$
22. $24c^4d^4$
23. -4
24. -A - 9B - 2
25. 3c + d
26. $-4R_0 + 3R + 3$
27. $-6x^3y^2 + 4x^3y^3 + 10x^2y$
28. $-7e^2f^5 + 9f^2 - 5$
29. m + 16
30. -2
31. -19
32. +4
33. -72
34. $9\dfrac{6}{7}$ ft
  or $118\dfrac{2}{7}$ in.

ANSWERS TO CHAPTER TESTS (continued)

FORM C

1. 16

2. 16

3a. $<$

3b. $>$

4. -11

5. -7

6. -5.6

7. -16

8. $\dfrac{5}{28}$

9. 54

10. $7R + 4S + 3$

11. $-5T_0 + 6T_1 - 7T_2$

12. $-70x^3y^5$

13. $-12a^2b^2 + 4ab^3$

14. $\dfrac{-2m^4}{v}$

15. $-5D^4 + 4D^2 + 1$

16. -14

17. $3 - 4j$

18. -13

19. -21

20. 147 cm^2

FORM D

1. 20

2. 7

3a. $>$

3b. $<$

4. -14

5. -1

6. -2

7. -5

8. -26

9. $\dfrac{1}{14}$

10. 14

11. 5

12. -40

13. 24

14. -22

15. $-\dfrac{1}{3}$

16. $-\dfrac{10}{9}$

17. a^{10}

18. $6b^3$

19. x^2

20. $5y^3$

21. $12a^4b^4$

22. 10

23. $2H - 7K + 13$

24. $10f + 4g - 4$

25. $M_0 + 5M - 2$

26. $-8d^3c^2 + 12d^3c^3 + 32d^2c$

27. $-3S^4 + 4RS^2 - \dfrac{6}{R}$

28. $3a + 10$

29. 15

30. $3b - 35$

31. 9

32. 126

33. 209 in. or $17\dfrac{5}{12}$ ft

4.1   Solve each equation for the given variable:

1. $x + 9 = 14$ 2. $R + 7 = -2$ 3. $\dfrac{e}{7} = -5$

4. $3T + 10 = -2$

4.2   5. $2Y + 19 = Y - 3$ 6. $3(j - 1) + 6 = j + 5$

7. $\dfrac{x}{16} + \dfrac{1}{8} = -\dfrac{3}{4}$

4.3   Solve the following inequalities and graph the solution on the number line:

8. $x - 4 > -2$ 9. $3Y \leqslant 15$ 10. $-7x \leqslant 21$

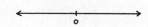   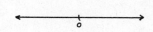   

11. $-\dfrac{e}{4} > -1$

4.4   Solve each of the following formulas for the indicated variable:

12. $r = \dfrac{I}{pt}$ for I   13. $r = \dfrac{I}{pt}$ for t

14. Use the formula from problem 13 to find t when $I = 40$, $P = 500$, and $r = 4\%$.

4.5   Translate each phrase into an algebraic expression. Let x represent the unknown number.

15. Twice the difference of a number and 10.

16. One third a number added to the number.

17. Translate "13 less than a number is -5" into an algebraic equation, and solve to find the unknown number.

4.6   Solve and check each of the following problems:

18. The width of a rectangle is 1 meter less than $\dfrac{2}{3}$ the length of the rectangle. Find the dimensions of the rectangle if the perimeter is 38 meters.

19. The sum of three angles of a triangle is 180°. The smaller angle is 20° less than the larger angle. The second angle is 10° less than the larger angle. Find the measure of each angle of the triangle.

4.7   20. Express 18 pounds to 24 pounds as a ratio in lowest terms.

21. Find x in the proportion: $\dfrac{x}{20} = \dfrac{7}{4}$

22. Translate "5 is to x as 2 is to 7" and solve for x.

4.8   23. If y varies directly as x, and $y = 90$ when $x = 15$, find y when $x = \dfrac{3}{2}$.

24. If y varies directly as x and z, and $y = 70$ when $x = 2$ and $y = 7$, find y when $x = \dfrac{1}{10}$ and $z = 40$.

25. If y varies inversely as x, and $y = 20$ when $x = 10$, find the constant of variation.

For problems 1 through 12, solve for the given variable:

1. $A + 9 = -11$

2. $-6 + r = 4$

3. $7 - T = -1$

4. $\dfrac{M}{2} = -6$

5. $-7T = 2$

6. $4b - 6 = 18$

7. $5(3x - 1) = -2x + 1$

8. $-2d + 8 = 6d - 32$

9. $\dfrac{y}{6} - 2y = \dfrac{3}{4}$

10. $\dfrac{R}{5} - \dfrac{2R - 5}{10} = \dfrac{R + 1}{5}$

11. $6(3f - 1) = -2(7 - 5f)$

12. $4 + .004x = 1.2$

13a. Solve for W: $S = \dfrac{3L(W - D)}{8}$

  b. Find W, given s = 3, L = 2, and D = 5.

14. Translate into algebraic language and solve: two-thirds an unknown decreased by 2 is 6.

15. An 18-ft board is cut into three pieces. If the first piece is three times the length of the second piece, and the third piece is two feet shorter than the second piece, find the length of the three pieces.

16. Inspectors for an electronic company rejected 120 discs. If this represented ½ percent of the daily production, how many discs were produced that day?

17. Express as a ratio in lowest form:

    a. 14 grams to 98 grams   b. 46 meters to .138 kilometers

18. If y varies directly as x and z, and y = 48 when x = 2 and z = 3, find y when x = 5 and z = 6.

For problems 19 and 20, solve the following inequalities and graph the solutions on the real number line:

19. $4x - 5 \geqslant 3$

20. $3(2x - 5) > 7(x - 1)$

1._____

2._____

3._____

4._____

5._____

6._____

7._____

8._____

9._____

10._____

11._____

12._____

13.a._____

13.b._____

14._____

15._____

16._____

17._____

18._____

19._____

20._____

CHAPTER 4 TEST
FORM B

For problems 1 through 12, solve for the given variable:

1. $H - 8 = -11$

2. $6 - T = 16$

3. $-5b = 45$

4. $\dfrac{k}{7} = -3$

5. $\dfrac{-2}{3} T_0 = 4$

6. $3x + 1 = -8$

7. $\dfrac{P}{6} - 2 = 1$

8. $7a - 2 = 6a + 4$

9. $-2(3x - 6) = 9$

10. $\dfrac{x}{4} + \dfrac{3x + 2}{12} = \dfrac{2x + 5}{3}$

11. $8.5 + .25x = 2.5$

12. $\dfrac{5K + 3}{6} - \dfrac{7K - 3}{4} = -\dfrac{3}{2}$

13a. Solve for D: $S = \dfrac{3L(W - D)}{8}$

b. Find D, given: $s = 0.6$, $L = 4.5$ and $W = 2.1$.

14. Translate into algebraic language and solve: One eighth of an unknown decreased by 2 is 9.

15. A triangle has a perimeter of 14 cm. One side is 1.4 cm longer than the shortest side and 4.3 cm shorter than the longest side. Find the dimensions of the triangle.

16. The mechanical advantage (MA) of any machine is the ratio of the load moved to effort applied. What is the MA of a system in which motion of 952 kg load is started by 28 kg of effort?

17. Express as a ratio in lowest form:

    a. 26 liters to 234 liters   b. 64 grams to .320 kilograms

18. If y varies inversely as x, and $y = 96$ when $x = 24$, find the value of y when $x = 16$.

For problems 19 and 20, solve the following inequalities and graph the solution on the real number line:

19. $2x - 7 \leq 5$

20. $2(3x + 1) < 8(x - 2)$

1. _____

2. _____

3. _____

4. _____

5. _____

6. _____

7. _____

8. _____

9. _____

10. _____

11. _____

12. _____

13a. _____

13.b. _____

14. _____

15. _____

16. _____

17.a. _____

17.b. _____

18. _____

19. _____

20. _____

For problems 1 through 11, solve for the given variable:

1. $D - 6 = -4$

2. $A + 5 = -7$

3. $-5b = 35$

4. $\dfrac{a}{6} = -1$

5. $-\dfrac{2}{3} T_0 = 8$

6. $3x + 1 = -7$

7. $\dfrac{P}{4} - 2 = 1$

8. $7a - 5 = 6a + 4$

9. $-3(2x - 5) = -3$

10. $\dfrac{x}{3} + \dfrac{3x + 2}{12} = \dfrac{2x + 5}{4}$

11. $\dfrac{k + 3}{6} - \dfrac{k - 3}{4} = -\dfrac{7}{2}$

12. Solve the formula $Q = P(Q_2 - Q_1)$ for Q_2.

13. Translate the following statement into algebraic language and solve: five times an unknown number increased by 18 is 78.

14. Express as a ratio: 7 meters to 11 meters.

15. Solve the given proportion for R:

$$\frac{7}{R} = \frac{49}{56}$$

16. If 9 grams of water will yield 1 gram of hydrogen, how many grams of hydrogen are yielded by 144 grams of water?

17. If y varies inversely as x, and $y = \frac{1}{4}$ when $x = 120$, find the value of y when $x = -3$.

18. The power dissipated in a circuit varies directly as the square of the voltage. What is the power dissipated if the constant of proportionality is .21 and the voltage is 4?

19. Graph the solution of $x \geq -1$ on the real number line.

20. Solve $3x - 5 > 7x + 3$ for the set of real numbers.

1. _____

2. _____

3. _____

4. _____

5. _____

6. _____

7. _____

8. _____

9. _____

10. _____

11. _____

12. _____

13. _____

14. _____

15. _____

16. _____

17. _____

18. _____

19. _____

20. _____

For problems 1 through 11, solve for the given variable:

1. $B + 6 = -8$

2. $-4 + t = 7$

3. $6 - R = -2$

4. $\dfrac{p}{4} = -3$

5. $-6x = 54$

6. $\dfrac{3}{4} y = -9$

7. $\dfrac{7}{8} = -7a$

8. $4L_0 + 6 = 3L_0 + 1$

9. $3(3x - 1) = 4x + 3$

10. $\dfrac{G}{5} - 2G = \dfrac{G + 19}{10}$

11. $\dfrac{R}{6} - \dfrac{3R - 2}{3} = \dfrac{7}{4}$

12. Solve for ℓ: $P - 2\ell + 2w$

13a. Solve for L: $M = \dfrac{Lt + q}{t}$

13b. Find L if $M = 20$, $t = 2$, and $g = 18$.

14. Two-fifths an unknown number is twelve less than the unknown number. Find the number.

15. A 60 kilogram weight is placed on one side of a fulcrum, and a second weight of 80 kg is placed on the other side. The 60 kg weight is placed 2 meters farther from the fulcrum than the 80 kg weight. How far is each weight from the fulcrum if the lever is balanced?

16. How many kilograms of candy worth 75 cents a kilogram should be mixed with 32 kilograms of candy worth 68 cents a kilogram to obtain a mixture worth 71 cents a kilogram?

17. Express as a ratio in lowest form:

    a. 8 liters to 28 liters   b. 42 meters to 4800 centimeters

18. Translate into a proportion and solve:

    x is to 6 as 5 is to 3

19. Translate into a proportion and solve: seventy-six liters of gas are used on a 616-kilometer trip. How many liters are needed for a trip of 1386 kilometers?

20. Write the given statement as an equation: the perimeter of a square varies directly as the length of the side.

21. If y varies directly as x and z, and $y = 30$ when $x = 6$ and $z = 10$, find y when $x = 14$ and $y = 3$.

22. Graph the inequality $x \leq -2$ on the real number line.

23. Solve for x: $-3x > 6$.

24. Solve for x: $2x - 2 \leq 4x - 2$.

1._____

2._____

3._____

4._____

5._____

6._____

7._____

8._____

9._____

10._____

11._____

12._____

13a._____

13b._____

14._____

15._____

16._____

17a._____

17b._____

18._____

19._____

20._____

21._____

22._____

23._____

24._____

ANSWERS TO CHAPTER 4 TESTS

PRETEST

1. X = 5

2. R = -9

3. e = -35

4. T = -4

5. Y = -22

6. j = 1

7. X = -14

8. X > 2

9. Y ≤ 5

10. X ≥ -3

11. e < 4

12. I = prt

13. $t = \dfrac{I}{pr}$

14. t = 2

15. 2(X - 10)

16. $X + \dfrac{1}{3}X$

17. X - 13 = -5,   X = 8

18. 7m by 12m

19. 50°, 60°, 70°

20. $\dfrac{3}{4}$

21. 35

22. X = 17.5

23. 9

24. 20

25. 200

FORM A

1. A = -20

2. r = 10

3. T = 8

4. M = -12

5. T = -2/7

6. b = 6

7. x = 6/17

8. d = 5

9. y = -9/22

10. R = 3/2

11. f = -1

12. x = -700

13a. $W = \dfrac{8s + 3LD}{3L}$ or $\dfrac{8s}{3L} + D$

  b. W = 9

14. 2/3n - 2 = 6, n = 12

15. 12 ft, 4 ft, and 2 ft

16. 24,000

17a. 1/7

  b. 1/3

18. 240

19.

20.

86

ANSWERS TO CHAPTER 4 TESTS (continued)

<u>FORM B</u>

1. $H = -3$

2. $T = -10$

3. $b = -9$

4. $k = -21$

5. $T_0 = -6$

6. $x = -3$

7. $p = 18$

8. $a = 6$

9. $x = \frac{1}{2}$

10. $x = -9$

11. $x = -24$

12. $K = 3$

13a. $D = \dfrac{3LW - 8s}{3L}$

  b. 1.7

14. $\frac{1}{8} n - 2 = 9$, $x = 88$

15. 3.7 cm by 2.3 cm by 8.0 cm

16. 34

17a. $\frac{1}{9}$

  b. $\frac{1}{5}$

18. 144

19. $x \leq 6$

20. $x \geq 9$

<u>FORM C</u>

1. $D = 2$

2. $A = -12$

3. $b = -7$

4. $a = -6$

5. $T_0 = -12$

6. $x = -8/3$

7. $p = 12$

8. $a = 9$

9. $x = 3$

10. $x = 13$

11. $k = 57$

12. $Q_2 = \dfrac{Q + PQ_1}{p}$

13. $5n + 18 = 78$, $n = 12$

14. $\dfrac{7}{11}$

15. $R = 8$

16. 16 grams

17. $y = -10$

18. 3.36

19.

20.

ANSWERS TO CHAPTER 4 TESTS (continued)

FORM D

1. $B = -14$

2. $t = 11$

3. $R = 8$

4. $p = -12$

5. $x = -9$

6. $y = -12$

7. $a = -1/8$

8. $L_0 = -5$

9. $x = \dfrac{6}{5}$

10. $G = -1$

11. $R = -13/10$

12. $\ell = \dfrac{P - 2w}{2}$

13a. $L = \dfrac{Mt - q}{t}$

  b. $L = 11$

14. $\dfrac{2}{5} n + 12 = n, \quad n = 20$

15. 6 m and 8 m

16. 24 kg

17a. $\dfrac{2}{7}$

  b. $\dfrac{7}{8}$

18. $x = 10$

19. 171 liters

20. $P = 4s$

21. $y = 21$

22.

23. $x < -2$

24. $x \geqslant 0$

CHAPTER 5 PRETEST

5.1 In problems 1 through 4, factor:

   1. $9A - 15$ 2. $PT + MT$

   3. $18x^3y^7 - 9x^2y^6 + 27x^4y^9$ 4. $a(b + 5) - 3(b + 5)$

5.2 In problems 5 through 8, multiply:

   5. $(x - 2)(x - 11)$ 6. $(w + 7)(w - 5)$

   7. $(4d + 3e)(5d + 3e)$ 8. $(2m + 5)(2m - 5)$

5.3 In problems 9 through 12, factor completely:

   9. $y^2 - 11y + 28$ 10. $A^2 + 4A + 3$

  11. $f^2 - 3f - 28$ 12. $4r^2 - 40r + 36$

5.4 In problems 13 through 16, factor completely:

  13. $6d^2 + 23d + 20$ 14. $42x^2 - 53x + 15$

  15. $15R^2 + 46RT + 16T^2$ 16. $18a^2 + 3a - 105$

5.5 In problems 17 and 18, square the following binomials:

  17. $(x - 9)^2$ 18. $(3e + 11)^2$

  In problems 19 through 21, factor completely:

  19. $B^2 - 24B + 144$ 20. $49y^2 - 25$ 21. $20e^2 - 60e + 45$

5.6 In problems 22 through 25, solve for the given variable:

  22. $x^2 - 10x + 9 = 0$ 23. $6R^2 + 14R = 0$

  24. $6T^2 + 11T - 121 = 0$ 25. $6C^2 + 21C - 216 = 0$

CHAPTER 5 TEST
FORM A

In problems 1 through 4, multiply:

1. $(3H - 5)(3H + 5)$

2. $(a - 9)(a + 2)$

3. $(x - 3y)^2$

4. $(4b - 3c)(3b - 4c)$

In problems 5 through 20, factor:

5. $15T - 24U$

6. $W^2 - 121$

7. $e^3f - ef^2$

8. $m^2 - m - 12$

9. $d^2 - 15d + 56$

10. $4B^2 - 4$

11. $42A^2 - A - 30$

12. $2M^2 + 14M + 24$

13. $4B^2 - 28B + 49$

14. $6n^2 + 24n$

15. $3a + ab + 3a^2 - ac$

16. $2(x + y) + z(x + y)$

17. $4x^2 + 16$

18. $2R^2 - 5R - 12$

19. $3T^2 - 9T - 30$

20. $24R^2 - 2R - 15$

In problems 21 through 25, solve for the given variable:

21. $x^2 - 11x + 10 = 0$

22. $2y^2 + 7y = 0$

23. $9a^2 - 6a + 1 = 0$

24. $2m^2 - 10m - 72 = 0$

25. $9R^2 - 1 = 0$

1._____

2._____

3._____

4._____

5._____

6._____

7._____

8._____

9._____

10._____

11._____

12._____

13._____

14._____

15._____

16._____

17._____

18._____

19._____

20._____

21._____

22._____

23._____

24._____

25._____

CHAPTER 5 TEST
FORM B

In problems 1 through 5, find the products:

1. $(a + 5b)(a - 5b)$ 1._____

2. $(11x - 2)(3x + 5)$ 2._____

3. $(7d + 9e)(6d + 5r)$ 3._____

4. $(4R - 3s)(7R - 2s)$ 4._____

5. $(7g + 3)^2$ 5._____

In problems 6 through 20, factor the following polynomials completely:

6. $81 - r^2$ 6._____

7. $y^2 - 15y + 44$ 7._____

8. $m^2n^2 - n^2$ 8._____

9. $4A^2 + 36A + 81$ 9._____

10. $H(D + 1) - E(D + 1)$ 10._____

11. $3a(2 + 5b) - 2(2 + 5b)$ 11._____

12. $3s^2 + 25s + 38$ 12._____

13. $2Z^2 + 4Z - 48$ 13._____

14. $W^3U^2 - 2W^2U^3 + 4U^4$ 14._____

15. $18m^2 + 21m - 15$ 15._____

16. $3y^2 + 27$ 16._____

17. $y(x - 3) + 4(x - 3)$ 17._____

18. $0.01R^2 + 0.02R - 0.15$ 18._____

19. $\frac{1}{3} x + \frac{1}{2} y + \frac{1}{8}$ 19._____

20. $8M^2 + 14MN - 15N^2$ 20._____

In problems 21 through 25, solve for the given variable:

21. $x^2 - 12x - 13 = 0$ 21._____

22. $3y^2 - 2y = 0$ 22._____

23. $4a^2 - 12a + 9 = 0$ 23._____

24. $3m^2 - 9m - 120 = 0$ 24._____

25. $4R^2 - 9 = 0$ 25._____

CHAPTER 5 TEST
FORM C

In problems 1 through 5, find the products:

1. $(5x + 9y)(5x - 9y)$

2. $(H - 2)(H + 3)$

3. $(2d - 5)(3d + 4)$

4. $(4a - 3)(3a - 1)$

5. $(6k - 7)^2$

In problems 6 through 20, factor completely:

6. $12a - 15b - 33$

7. $2m^2n + 3mn^2 - 4mn$

8. $42R^3S^2 - 48R^5S^3 - 66R^4S^3 + 54R^3S^3$

9. $4x^2 - 12xy + 9y^2$

10. $25D^2 - 1$

11. $p^2 - 5p - 14$

12. $r^2 - 5r + 6$

13. $12c^2 + 12c - 9$

14. $42e^2 + 71e + 30$

15. $a(x + 7) - b(x + 7)$

16. $8s^2 - 18$

17. $20T_0^2 - T_0 - 63$

18. $a^4 - 8a^3 + 16a^2$

19. $72d^2 + 87de + 21e^2$

20. $6x^2 + 6$

In problems 21 through 25, solve for the given variable:

21. $x^2 - 16x + 15 = 0$

22. $5y^2 - 3y = 0$

23. $25a^2 + 10a + 1 = 0$

24. $3m^2 + 9m - 120 = 0$

25. $9R^2 - 4 = 0$

1._____

2._____

3._____

4._____

5._____

6._____

7._____

8._____

9._____

10._____

11._____

12._____

13._____

14._____

15._____

16._____

17._____

18._____

19._____

20._____

21._____

22._____

23._____

24._____

25._____

CHAPTER 5 TEST
FORM D

In problems 1 through 6, find the products:

1. $(x - 2)(x - 5)$

2. $(2H + 7)(2H + 5)$

3. $(3W + 7x)(3W - 7x)$

4. $(a - 11)(a + 4)$

5. $(\frac{1}{4}B + \frac{1}{3})(\frac{1}{3}B + \frac{1}{2})$

6. $(2P + 9)^2$

In problems 7 through 20, factor completely:

7. $32y^2 - 48y^3 + 80y$

8. $\frac{1}{4}b + \frac{1}{3}c + \frac{1}{2}$

9. $0.2f^3 - 0.04f^2 + 0.4f^5$

10. $64A^2 - 9$

11. $16r^2 - 24r + 9$

12. $F^2 - 4F - 5$

13. $2C^2 + 10C + 12$

14. $56E^2 + 93E + 27$

15. $d^4 - d^3 - 6d^2$

16. $5(r - s) + c(r - s)$

17. $25N^2 - 30N + 9$

18. $18Z^2 + 33Z - 30$

19. $36R^2 - 59RS + 24S^2$

20. $20 + 5x^2$

In problems 21 through 25, solve for the given variable:

21. $x^2 - 15x - 16 = 0$

22. $4y^2 - 5y = 0$

23. $9a^2 + 12a + 4 = 0$

24. $2m^2 - 6m - 80 = 0$

25. $9R^2 - 16 = 0$

1._____

2._____

3._____

4._____

5._____

6._____

7._____

8._____

9._____

10._____

11._____

12._____

13._____

14._____

15._____

16._____

17._____

18._____

19._____

20._____

21._____

22._____

23._____

24._____

25._____

ANSWERS TO CHAPTER 5 TESTS

PRETEST

1. $3(3A - 5)$

2. $T(P + M)$

3. $9X^2Y^6(2XY - 1 + 3X^2Y^3)$

4. $(b + 5)(a - 3)$

5. $X^2 - 13X + 22$

6. $W^2 + 2W - 35$

7. $20d^2 + 27de + 9e^2$

8. $4m^2 - 25$

9. $(Y - 4)(Y - 7)$

10. $(A + 1)(A + 3)$

11. $(f - 7)(f + 4)$

12. $4(r - 9)(r - 1)$

13. $(2d + 5)(3d + 4)$

14. $(7X - 3)(6X - 5)$

15. $(5R + 2T)(3R + 8T)$

16. $3(2a + 5)(3a - 7)$

17. $X^2 - 18X + 81$

18. $9e^2 + 66e + 121$

19. $(B - 12)^2$

20. $(7Y + 5)(7Y - 5)$

21. $5(2e - 3)^2$

22. $X = 1$ or $X = 9$

23. $R = 0$ or $R = -\frac{7}{3}$

24. $T = \frac{11}{3}$ or $T = -\frac{11}{2}$

25. $C = \frac{9}{2}$ or $C = -8$

FORM A

1. $9H^2 - 25$

2. $a^2 - 7a - 18$

3. $x^2 - 6xy + 9y^2$

4. $12b^2 - 25bc + 12c^2$

5. $3(5T - 8U)$

6. $(W - 11)(W + 11)$

7. $ef(e^2 - f)$

8. $(m - 4)(m + 3)$

9. $(d - 7)(d - 8)$

10. $4(B - 1)(B + 1)$

11. $(6A + 5)(7A - 6)$

12. $2(M + 3)(M + 4)$

13. $(2B - 7)^2$

14. $3n(2n + 8)$

15. $a(3 + b + 3a - c)$

16. $(x + y)(2 + z)$

17. $4(X^2 + 4)$

18. $(2R + 3)(R - 4)$

19. $3(T - 5)(T + 2)$

20. $(6r - 5)(4R + 3)$

21. $X = 1$ or $X = 10$

22. $Y = 0$ or $Y = -\frac{7}{2}$

23. $a = \frac{1}{3}$

24. $M = 9$ or $M = -4$

25. $R = \frac{1}{3}$ or $R = -\frac{1}{3}$

ANSWERS TO CHAPTER 5 TESTS (continued)

FORM B

1. $a^2 - 25b^2$

2. $33x^2 + 49x - 10$

3. $42d^2 + 89de + 45e^2$

4. $28R^2 - 29Rs + 6s^2$

5. $49g^2 + 42g + 9$

6. $(9 - r)(9 + r)$

7. $(y - 4)(y - 11)$

8. $n^2(m - 1)(m + 1)$

9. $(2A + 9)^2$

10. $(D + 1)(H - E)$

11. $(2 + 5b)(3a - 2)$

12. $(3s + 19)(s + 2)$

13. $2(Z - 4)(Z + 6)$

14. $U^2(W^3 - 2W^2U + 4U^2)$

15. $3(2m - 1)(3m + 5)$

16. $3(Y^2 + 9)$

17. $(x - 3)(y + 4)$

18. $0.01(R - 3)(R + 5)$

19. $\frac{1}{24}(8x + 12y + 3)$

20. $(4M - 3N)(2M + 5N)$

21. $X = 13$, or $X = -1$

22. $Y = 0$, or $Y = \frac{2}{3}$

23. $a = \frac{3}{2}$

24. $M = 8$, or $M = -5$

25. $R = \frac{3}{2}$, or $R = -\frac{3}{2}$

FORM C

1. $25x^2 - 81y^2$

2. $H^2 + H - 6$

3. $6d^2 - 7d - 20$

4. $12a^2 - 13a + 3$

5. $36k^2 - 84k + 49$

6. $3(4a - 5b - 11)$

7. $mn(2m + 3n - 4)$

8. $6R^3S^2(7 - 8R^2S - 11R + 9S)$

9. $(2x - 3y)^2$

10. $(5D - 1)(5D + 1)$

11. $(p - 7)(p + 2)$

12. $(r - 3)(r - 2)$

13. $3(2c - 1)(2c + 3)$

14. $(7e + 6)(6e + 5)$

15. $(a - b)(x + 7)$

16. $2(2s - 3)(2s + 3)$

17. $(5T_0 - 9)(4T_0 + 7)$

18. $a^2(a - 4)^2$

19. $3(3d + e)(8d + 7e)$

20. $6(X^2 + 1)$

21. $X = 1$ or $X = 15$

22. $Y = 0$ or $Y = \frac{3}{5}$

23. $a = -\frac{1}{5}$

24. $M = 5$ or $M = -8$

25. $R = \frac{2}{3}$ or $R = -\frac{2}{3}$

FORM D

1. $x^2 - 7x + 10$

2. $4H^2 + 24H + 35$

3. $9W^2 - 49X^2$

4. $a^2 - 7a - 44$

5. $\frac{1}{12} B^2 + \frac{17}{72} B^2 + \frac{1}{6}$

6. $4P^2 + 36P + 81$

7. $16y(2y - 3y^2 + 5)$

8. $\frac{1}{12} (3b + 4c + 6)$

9. $0.04f^2(5f - 1 + 10f^3)$

10. $(8A - 3)(8A + 3)$

11. $(4r - 3)^2$

12. $(F - 5)(F + 1)$

13. $2(C + 2)(C + 3)$

14. $(8E + 3)(7E + 9)$

15. $d^2(d - 3)(d + 2)$

16. $(r - s)(5 + c)$

17. $(5N - 3)^2$

18. $3(2Z + 5)(3Z - 2)$

19. $(9R - 8S)(4R - 3S)$

20. $5(4 + X^2)$

21. $X = -1$ or $X = 16$

22. $Y = 0$ or $Y = \frac{5}{4}$

23. $a = -\frac{2}{3}$

24. $M = -5$ or $M = 8$

25. $R = \frac{4}{3}$ or $R = -\frac{4}{3}$

6.1   Reduce to lowest terms:

1. $\dfrac{4a - 16}{a^2 - 16}$

2. $\dfrac{2x^2 - 3x - 9}{2x^2 + 11x + 12}$

6.2   Perform the indicated operations:

3. $\dfrac{x^2 - 9}{3} \cdot \dfrac{3x - 9}{x^2 + 7x - 12}$

4. $\dfrac{x^2 - 5x + 6}{x^2 - 4} \div \dfrac{x^2 - 8x + 15}{2x + 4}$

6.3   5. Find the L.C.D. of the following algebraic fractions and convert to equivalent fractions with a common denominator:

$$\dfrac{2}{x^2 - 25} , \quad \dfrac{5}{x^2 - x - 20}$$

6.4   Perform the indicated operations and simplify the answers:

6. $\dfrac{3}{x - y} - \dfrac{2}{x + y}$

7. $\dfrac{x}{x^2 - 9} + \dfrac{2}{x^2 - 6x + 9}$

6.5   Solve and check:

8. $\dfrac{3}{x - 2} + \dfrac{4}{x + 2} = \dfrac{5}{x^2 - 4}$

In problems 1 through 5, reduce to lowest terms:

1. $\dfrac{2R^2}{2R^2 - 6R}$

1._____

2. $\dfrac{5x - 5y}{-2x + 2y}$

2._____

3. $\dfrac{T^2 + T - 12}{2T^2 - 7T + 3}$

3._____

4. $\dfrac{3B^2 - 9B}{6B}$

4._____

5. $\dfrac{x^2 - 3x - 4}{x^2 - 7x + 12}$

5._____

In problems 6 through 20, perform the indicated operations:

6. Multiply: $\dfrac{4a}{a^2 - 4} \cdot \dfrac{3a - 6}{12a^2}$

6._____

7. Multiply: $\dfrac{A^2 + 3A}{A^2 - 9} \cdot \dfrac{A^2 - 4A + 3}{A^2 - 1}$

7._____

8. Divide: $\dfrac{(r - 4)^2}{3r} \div \dfrac{B^2 + 5B + 6}{B^2 + B - 2}$

8._____

9. Divide: $\dfrac{(B + 3)^2}{B - 1} \div \dfrac{r^2 - 16}{12r^2}$

9._____

10. Add: $\dfrac{2}{c^2 + c - 2} + \dfrac{3}{c^2 - c - 6}$

10._____

11. Subtract: $\dfrac{3e + 1}{10} - \dfrac{2e - 3}{15}$

11._____

12. Subtract: $\dfrac{x + 3}{x^2 - 1} - \dfrac{x - 2}{x^2 + 2x + 1}$

12._____

13. Subtract: $\dfrac{2f + 1}{8} - \dfrac{3f - 5}{6}$

13._____

14. Subtract: $\dfrac{H + 2}{H^2 - 1} - \dfrac{H + 1}{x^2 - 2H + 1}$

14._____

15. Subract: $\dfrac{y + 1}{2y - 8} - \dfrac{y - 1}{y^2 - y - 12}$

15._____

16. Solve: $\dfrac{b + 1}{4} - \dfrac{b - 2}{8} = \dfrac{5}{6}$

16._____

17. Solve: $\dfrac{x - 3}{x^2 - 1} + \dfrac{x + 1}{x^2 + x - 2} = \dfrac{2x + 5}{x^2 + 3x + 2}$

17._____

18. Solve: $\dfrac{1}{2x} = \dfrac{7}{6} - \dfrac{3}{x}$

18._____

19. Solve: $\dfrac{T + 5}{T^2 - 4} = \dfrac{T - 3}{T^2 + 4T + 4}$

19._____

20. Solve: $\dfrac{1}{3x} = \dfrac{7}{6} - \dfrac{2}{x}$

20._____

CHAPTER 6 TEST
FORM B

In problems 1 through 5, reduce to lowest terms:

1. $\dfrac{2D^2 - 4D}{4D}$

1._____

2. $\dfrac{a^2 - 4a - 5}{a^2 - 7a + 10}$

2._____

3. $\dfrac{6x}{4x^2 - 12x}$

3._____

4. $\dfrac{a^2 - 6a + 9}{a^2 - 9}$

4._____

5. $\dfrac{2b^2 - 2}{4b^2 - 16b + 12}$

5._____

In problems 6 through 20, perform the indicated operations:

6. $\dfrac{7M^2}{2N} \cdot \dfrac{5N^3}{14MN}$

6._____

7. $\dfrac{I^2 + 5I}{I^2 - 25} \cdot \dfrac{I^2 - 6I + 5}{I^2 - 1}$

7._____

8. $\dfrac{e^2 - 3e + 2}{3e^3 - 3e^2} \div \dfrac{e^2 - 3e + 2}{6e^2 - 6e}$

8._____

9. $\dfrac{2y^2 + 5y + 3}{2y^2 - 4y - 6} \div \dfrac{8y + 12}{y^2 + 2y - 15}$

9._____

10. $\dfrac{H - 1}{10} + \dfrac{H + 2}{15}$

10._____

11. $\dfrac{2f + 1}{8} - \dfrac{3f - 5}{6}$

11._____

12. $\dfrac{H + 2}{H^2 - 1} - \dfrac{H + 1}{H^2 - 2H + 1}$

12._____

13. $\dfrac{H + 2}{H^2 - 2H - 3} - \dfrac{H + 3}{H^2 - H - 2}$

13._____

14. $\dfrac{x - 1}{2x - 4} - \dfrac{2x^2}{4x^2 - 16}$

14._____

15. $\dfrac{6}{y + 3} + \dfrac{3}{y^2 - 9}$

15._____

16. $\dfrac{R - 2}{18} + \dfrac{R + 3}{54}$

16._____

17. $\dfrac{2y}{4} - \dfrac{6y}{5} = \dfrac{1}{10}$

17._____

18. $\dfrac{R + 1}{4R} - \dfrac{9}{2R} = -3$

18._____

In problems 19 and 20, solve for the given variable:

19._____

19. $\dfrac{1}{3x} = \dfrac{7}{6} - \dfrac{2}{x}$

20. $\dfrac{T - 2}{T^2 - 1} = \dfrac{T - 1}{T^2 + 2T + 1}$

20._____

99

CHAPTER 6 TEST
FORM C

In problems 1 through 5, reduce to lowest terms:

1. $\dfrac{8x}{4x + 12}$

2. $\dfrac{2H - 4H^2}{4H}$

3. $\dfrac{a^2 - 7a + 10}{a^2 - 4}$

4. $\dfrac{3d^2 + 15d}{3d^2 - 6d}$

5. $\dfrac{T^2 + 2T - 3}{T + 3}$

In problems 6 through 15, perform the indicated operations:

6. $\dfrac{5xy^2}{3z} \cdot \dfrac{12z^2}{15xy^3}$

7. $\dfrac{2T^2 - 2T}{T^2 - 1} \cdot \dfrac{T^2 + 4T + 3}{T^2 - 2T - 15}$

8. $\dfrac{b^2 - 3b + 2}{b^2 - 9} \cdot \dfrac{b^2 - 3b}{2b - 2} \cdot \dfrac{2b + 6}{b^2 + 2b}$

9. $\dfrac{3M}{m} \div \dfrac{M^2}{2m}$

10. $\dfrac{2j + 14}{4j - 16} \div \dfrac{j^2 + 9j + 14}{j^2 - 4j}$

11. $\dfrac{2W + 3}{8} + \dfrac{5W - 2}{6}$

12. $\dfrac{7e + 1}{3} - \dfrac{2e - 3}{7}$

13. $\dfrac{T - 2}{T + 4} + \dfrac{T + 3}{T - 1}$

14. $\dfrac{g + 3}{g - 2} - \dfrac{g - 3}{g + 1}$

15. $\dfrac{D - 2}{D^2 - 2D + 1} - \dfrac{D - 5}{D^2 + 2D - 3}$

In problems 16 through 20, solve for the given variable:

16. $\dfrac{3}{x} + \dfrac{2}{5} = 1$

17. $\dfrac{3}{8y} = \dfrac{1}{6y} - \dfrac{5}{4}$

18. $\dfrac{c + 1}{c - 4} = 1 - \dfrac{3}{c + 4}$

19. $\dfrac{x - 2}{x^2 + 5x} = \dfrac{3}{2x + 10} = \dfrac{2x - 1}{2x^2 + 10x}$

20. $\dfrac{d + 1}{d^2 - 4} = \dfrac{2d - 3}{d^2 + 5d + 6} - \dfrac{d + 1}{d^2 + d - 6}$

1._____

2._____

3._____

4._____

5._____

6._____

7._____

8._____

9._____

10._____

11._____

12._____

13._____

14._____

15._____

16._____

17._____

18._____

19._____

20._____

CHAPTER 6 TEST
FORM D

In problems 1 through 5, reduce to lowest terms:

1. $\dfrac{5y}{5y - 20}$

2. $\dfrac{i^2 - 4i}{4i}$

3. $\dfrac{R^2 + 5R + 6}{R^2 - 9}$

4. $\dfrac{A^2 + 6A + 9}{A + 3}$

5. $\dfrac{2e^3 - 10e^2}{4e^2 - 24e + 20}$

In problems 6 through 15, perform the indicated operations:

6. $\dfrac{7A^2B}{3C^2} \cdot \dfrac{33C^2}{14AB}$

7. $\dfrac{2x}{x + 4} \cdot \dfrac{x^2 - 16}{2x^2}$

8. $\dfrac{f^2 - 3f + 2}{f^2 - f - 2} \cdot \dfrac{f^2 + 4f + 3}{f^2 + 3f - 4}$

9. $\dfrac{2}{D} \div \dfrac{D}{160}$

10. $\dfrac{3H^2 + 15H}{H^2 + 3H} \div \dfrac{H^2 + H}{3H + 9}$

11. $\dfrac{5r - 1}{6} + \dfrac{3r + 1}{4}$

12. $\dfrac{4t - 2}{15} - \dfrac{3t - 1}{10}$

13. $\dfrac{a - 3}{a^2 + 4a + 3} + \dfrac{a + 2}{a^2 - 9}$

14. $\dfrac{T - 1}{T^2 - 5T} - \dfrac{T - 2}{T^2 - 2T - 15}$

15. $\dfrac{2j}{j^2 - j - 12} + \dfrac{j + 5}{j^2 + 2j - 3} - \dfrac{j + 3}{j^2 - 5j + 4}$

In problems 16 through 20, solve for the given variable:

16. $\dfrac{2}{W} + \dfrac{7}{5} = 1$

17. $\dfrac{y - 1}{9y} = \dfrac{y + 1}{12y} - \dfrac{5}{6y}$

18. $\dfrac{Z - 2}{Z - 1} = \dfrac{2Z - 3}{Z + 1} - 1$

19. $\dfrac{b - 4}{b^2 + 3b} - \dfrac{5}{2b + 6} = \dfrac{3}{2b^2 + 6b}$

20. $\dfrac{2x + 1}{x^2 + 2x - 15} = \dfrac{3x + 1}{x^2 + 6x + 5} - \dfrac{x + 5}{x^2 - 2x - 3}$

1._____

2._____

3._____

4._____

5._____

6._____

7._____

8._____

9._____

10._____

11._____

12._____

13._____

14._____

15._____

16._____

17._____

18._____

19._____

20._____

ANSWERS TO CHAPTER 6 TESTS

PRETEST

1. $\dfrac{4}{a + 4}$

2. $\dfrac{x - 3}{x + 4}$

3. $\dfrac{(x - 3)^2}{x + 4}$

4. $\dfrac{1}{x - 5}$

5. L.C.D. $= (x + 5)(x - 5)(x + 4)$;

$\dfrac{2x + 8}{(x + 5)(x - 5)(x + 4)}$; $\dfrac{5x + 25}{(x + 5)(x - 5)(x + 4)}$

6. $\dfrac{x + 5y}{(x - y)(x + y)}$

7. $\dfrac{x^2 - x + 6}{(x + 3)(x - 3)^2}$

8. $x = 1$

FORM A

1. $\dfrac{R}{R - 3}$

2. $-\dfrac{5}{2}$

3. $\dfrac{T + 4}{2T - 1}$

4. $\dfrac{B - 3}{2}$

5. $\dfrac{x + 1}{x - 3}$

6. $\dfrac{1}{a(a + 2)}$

7. $\dfrac{A}{A + 1}$

8. $\dfrac{B - 1}{B + 2}$

9. $B + 3$

10. $\dfrac{5c - 9}{(c + 2)(c - 1)(c - 3)}$

11. $\dfrac{5e + 9}{30}$

12. $\dfrac{7x + 1}{(x - 1)(x + 1)^2}$

13. $\dfrac{-6f + 23}{24}$

14. $\dfrac{-H - 3}{(H - 1)^2(H + 1)}$

15. $\dfrac{y^2 + 2y + 5}{2(y - 4)(y + 3)}$

16. $b = 8/3$

17. $x = 0$

18. $x = 3$

19. $T = -1/3$

20. $x = 2$

ANSWERS TO CHAPTER 6 TESTS (continued)

FORM B

1. $\dfrac{D - 2}{2}$

2. $\dfrac{a + 1}{a - 2}$

3. $\dfrac{3}{2(x - 3)}$

4. $\dfrac{a - 3}{a + 3}$

5. $\dfrac{b + 1}{2(b - 3)}$

6. $\dfrac{5MN}{4}$

7. $\dfrac{I}{I + 1}$

8. $\dfrac{2}{e}$

9. $\dfrac{y + 5}{8}$

10. $\dfrac{5H + 1}{30}$

11. $\dfrac{-6f + 23}{24}$

12. $\dfrac{-H - 3}{(H - 1)^2(H + 1)}$

13. $\dfrac{5}{(H - 3)(H + 1)(H - 2)}$

14. $\dfrac{1}{2(x + 2)}$

15. $\dfrac{6y - 15}{(y + 3)(y - 3)}$

16. $\dfrac{4R - 3}{54}$

17. $y = -\dfrac{1}{7}$

18. $R = \dfrac{17}{33}$

19. $x = 2$

20. $T = 3$

FORM C

1. $\dfrac{2x}{x + 3}$

2. $\dfrac{1 - 2H}{2}$

3. $\dfrac{a - 5}{a + 2}$

4. $\dfrac{d + 5}{d - 2}$

5. $T - 1$

6. $\dfrac{4z}{3y}$

7. $\dfrac{2T}{T - 5}$

8. $\dfrac{b - 2}{b + 2}$

9. $\dfrac{6}{M}$

10. $\dfrac{j}{2(j + 2)}$

11. $\dfrac{26W + 1}{24}$

12. $\dfrac{43e + 16}{21}$

13. $\dfrac{2T^2 + 4T + 14}{(T + 4)(T - 1)}$

14. $\dfrac{9g - 3}{(g - 2)(g + 1)}$

15. $\dfrac{7D - 11}{(D - 1)^2(D + 3)}$

16. $x = 5$

17. $y = -1/6$

18. $c = -1$

19. $x = -1$

20. $d = 1/14$

FORM D

1. $\dfrac{y}{y - 4}$

2. $\dfrac{i - 4}{4}$

3. $\dfrac{R + 2}{R - 3}$

4. $A + 3$

5. $\dfrac{e^2}{2(e - 1)}$

6. $\dfrac{11A}{2}$

7. $\dfrac{x - 4}{x}$

8. $\dfrac{f + 3}{f + 4}$

9. $\dfrac{320}{D^2}$

10. $\dfrac{9(H + 5)}{H(H + 1)}$

11. $\dfrac{19r + 1}{12}$

12. $\dfrac{-t - 1}{30}$

13. $\dfrac{2a^2 - 3a + 11}{(a + 1)(a + 3)(a - 3)}$

14. $\dfrac{4T - 3}{T(T - 5)(T + 3)}$

15. $\dfrac{2j^2 - 7j - 29}{(j - 4)(j + 3)(j - 1)}$

16. $W = -5$

17. $y = -23$

18. $Z = 3/2$

19. $b = -10/3$

20. $x = -\dfrac{21}{29}$

7.1   1. Mrs. Stern taught five sections of Introductory Biology.
          Out of a total of 120 students in the five classes, grades
          at the end of the semester were as follows: 12 A's, 18 B's,
          48 C's, and 18 D's. How many students failed the course?
          Construct a circle graph to illustrate the grade distribution.

      2. Construct a bar graph to illustrate the value of 100 shares
          of PSD Stock during a six-month time period: January 31 – $2000;
          February 28 – $1600; March 31 – $2200; April 30 – $2500; May 31 –
          $2800; June 30 – $2600.

      3. Construct a line graph to illustrate the information in problem 2.

7.2   Draw the geometric figure given the coordinates of its vertices, and
      identify the figure.

      4. (1, 3), (-3, 2), (2, 1)

      5. (4, 6), (6, 6), (2, 3), (8, 3)

      State the domain and range of the given relation, and state whether
      the relation is a function.

      6. $\{(1, 8), (2, 8), (3, 8)\}$

      7. $\{(3, 2), (3, 5)\}$

      Give the quadrant of the given point.

      8. (-3, -7)

      9. (4, -1)

7.3   Give the slope and y-intercept of the given linear equation.

      10. $y = -3x + 5$

      11. $2x - y = 7$

      Graph the given linear equations by finding ordered pairs that are solutions
      to the given equations.

      12. $y = -x + 3$

      13. $2x + y = -1$

      14. $6x - 3y = 2$

7.4   Graph the given linear equations using the slope-intercept method.

      15. $y = x - 2$ 16. $5x + y = -3$ 17. $12x - 4y = -8$

      Find the slope of the line that passes through the given points.

      18. (4, -1), (2, 3)   19. (-5, -1), (-8, -6) 20. (-3, -1), (8, -1)

1. Draw a bar graph to illustrate the mileage of the following
   vehicles:

Chevrolet	18 mpg
Datsun	23 mpg
Toyota	25 mpg
GMAC	17 mpg
International	14mpg

2. Construct a circle graph to illustrate how each yearly overhead
   was allocated:

Rental	30,000
Utilities	10,000
Licenses	15,000
Salaries	20,000
Insurance	5,000

3. Construct a line graph to illustrate the earnings of a firm over
   a five-year period:

1980	500,000
1981	750,000
1982	675,000
1983	800,000
1984	850,000

4. Draw a geometric figure given the coordinates of the vertices:
   (2, 5), (-3, 5), (1, 2), and (-4, 2), and tell what figure it is.

5. Draw the geometric figure given the coordinates of its vertices,
   and identify the figure: (-3, 1), (1, 3), (3, -2)

   Graph the given equations:

6. $2x + y = 6$ 7. $x - y = -3$ 8. $y - 7 = 0$

9. $x + y = 0$ 10. $x = 2$

Tell whether the following are functions. If a function, give the domain and range.

11. $\{(3, 1), (2, 4), (5, 1)\}$ 12. $\{(3, 5), (5, 3), (6, 7), (3, 4)\}$

13. $y - 6 = 5$

Give the quadrant of the given point:

14. (-2, 3)   15. (1, 1)

Give the slope and y-intercept of the given linear equation.

16. $y = 7x - 1$ 17. $2x - 3y = 5$

Find the slope of the line that passes through the given points.

18. (8, 9), (7, 6)   19. (-2, 4), (-3, -3)   20. (4, 7), (-2, 3)

CHAPTER 7 TEST
FORM B

1. Draw a bar graph to illustrate the contents of an electrician's truck:

1/2" PVC	100 ft.
1' PVC	120 ft.
1/2" EMT	150 ft.
1" EMT	90 ft.
1/2" GRC	80 ft.
1" GRC	50 ft.

2. Construct a circle graph to illustrate how yearly earnings from a construction firm were allocated:

Labor	400,000
Material	450,000
Equipment Rental	80,000
Overhead	120,000
Profit	100,000

3. Construct a line graph to illustrate the number of employees of a firm over a five-year period:

1980	15
1981	23
1982	20
1983	18
1984	22

4. Draw a geometric figure given the coordinates of the vertices, and tell what kind of figure it is: (2, -1), (2, 4), (-3, -1)

5. Draw a geometric figure given the coordinates of the vertices, and tell what kind of figure it is: (1, 1), (1, 3), (3, 1), (3, 3)

In problems 6 through 10, graph the given equations:

6. $3y = 2x$

7. $x + y + 2 = 0$

8. $5x + y = 5$

9. $y - 6 = -4$

10. $x + 3 = 0$

In problems 11 through 13, tell whether the following relations are functions. If functions, state the domain and the range.

11. $\{(6, 5), (5, 6)\}$

12. $\{(1, 4), (1, -4)\}$

13. $x = 4$

In problems 14 and 15, give the quadrant of the given point:

14. $(-19, -1)$

15. $(-19, 3)$

In problems 16 and 17, give the slope and y-intercept of the given linear equation.

16. $y = -2x + 11$

17. $4x + 9y = -3$

In problems 18 through 20, find the slope of the line that passes through the given points.

18. $(16, 6), (17, 3)$

19. $(1, 3), (-2, 3)$

20. $(-10, -2), (4, -6)$

CHAPTER 7 TEST
FORM C

1. Draw a bar graph to illustrate the resistivity of the given materials in ohms
    per mil-foot at 20° C.

Platinum-iridium	148
Monel metal	253
Iala	301
Superior	520
Nichrome	600
Calorite	720

2. Construct a circle graph to illustrate the annual average of job openings in the
    field of Natural Science as given by the following information:

Geologists	800
Geophysicists	300
Meterologists	200
Oceanographers	500

3. Construct a line graph to illustrate the sales of a lumber company during a
    three-year period.

6-month period	Sales
1	20,000
2	40,000
3	60,000
4	26,000
5	80,000
6	46,000

In problems 4 and 5, draw a geometric figure given the coordinates of the vertices,
and tell what the figure is.

4. (-5, 4), (-5, -3), (-3, -3), (-3, 4)   5. (0, 3), (-2, 0), (5, 0)

In problems 6 through 10, graph the given equations on the number plane:

6. $x + y = 7$ 7. $3x - y = -2$ 8. $x - y = 0$

9. $y + 5 = 0$ 10. $x - 5 = 0$

In problems 11 through 13, state the domain and the range of the following relations,
and tell whether the relation is a function:

11. (4, 2), (3, 2), (1, 2)   12. (2, 1), (1, 2), (2, 3), (5, 6)

13. $y - 4 = 2$

In problems 14 and 15, give the quadrant of the given point:

14. (6, 1)   15. (6, -1)

In problems 16 and 17, give the slope and y-intercept of the given linear equation:

16. $y = -8x + 5$ 17. $x + 5y = 5$

In problems 18 through 20, find the slope of the line that passes through the
given points:

18. (-23, 14), (-24, 17)   19. (3, -4), (12, -4)   20. (5, -5), (-2, -2)

1. Draw a bar graph to illustrate the sidereal period of the revolution around the sun of the following planets:

   Pluto   247 years
   Neptune   164 years
   Jupiter   29 years
   Uranus   84 years
   Saturn   29 years

2. Draw a circle graph to illustrate the population of the four largest countries in the world.

   China   745,000,000
   India   543,000,000
   Soviet Union   241,000,000
   United States   204,300,000

3. Draw a line graph to illustrate the highest point in the states bordering the Gulf of Mexico.

   Texas   8,751 feet
   Louisiana   535 feet
   Mississippi   806 feet
   Alabama   2,407 feet
   Florida   345 feet

4. Draw the geometric figure given the coordinates of the vertices (3, 4), (-5, -1), and (1, -6). Tell which figure it is.

In problems 5 through 8, graph the given equations on the number plane:

5. $y = 4x$

6. $x + y - 1 = 0$

7. $2x + y = 4$

8. $y - 8 = 1$

In problems 9 through 11, state the domain and the range of the following relations, and tell whether the relation is a function.

9. (6, 8), (8, 8)

10. (7, 1), (-1, 7), (2, 3), (-3, 2)

11. $x - 4 = 1$

12. Draw the geometric figure given the coordinates of the vertices: (1, 2), (0, -1), (4, 2), and (5, -1).

13. Find the area of the figure in problem 12 if each unit represents one meter.

In problems 14 and 15, give the quadrant of the given point:

14. (-1, 80)

15. (100, 100)

In problems 16 and 17, give the slope and y-intercept of the given linear equation:

16. $y = 11x - 11$

17. $4x - 2y = 40$

In problems 18 through 20, find the slope of the line that passes through the given points:

18. (14, -5), (13, 11)

19. (-20, 40), (10, 50)

20. (8, 3), (8, 5)

ANSWERS TO CHAPTER 7 TESTS

PRETEST

1. 24 students failed

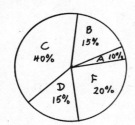

2.

```
JUNE 30  ████████████
MAY 31   █████████████
APRIL 30 ███████████
MAR 31   ██████████
FEB 28   █████████
JAN 31   █████████
         $1000  $2000  $3000
          STOCK VALUE
```

3.
```
$3000
$2000
$1000
      JAN 31  FEB 28  MAR 31  APR 30  MAY 31  JUNE 30
```

4. triangle

(-3,2) (1,3) (2,1)

5. trapezoid

(4,6) (6,6) (2,3) (8,3)

6. Domain: $\{1, 2, 3\}$
   Range: $\{8\}$, function

7. Domain: $\{3\}$
   Range: $\{2, 5\}$, not a function

8. III

9. IV

10. -3, 5

11. 2, -7

12.

(0,3) (3,0)

13.

(-1,1) (0,-1)

14.

$(\frac{1}{5}, 0)$ $(0, -\frac{2}{5})$

15.

110

<u>PRETEST</u> (continued)

16.

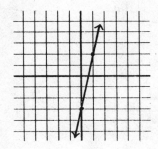

17.

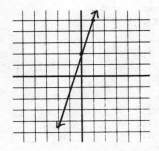

18. -2

19. $\dfrac{5}{3}$

20. 0

FORM A

1.

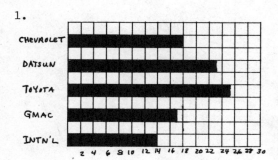

2.

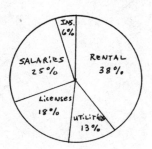

3.

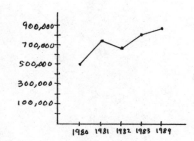

4. parallelogram

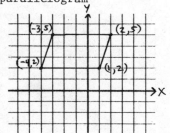

5. triangle

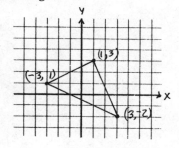

6.

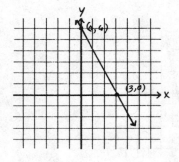

7.

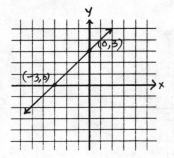

8.

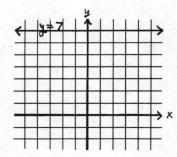

9.

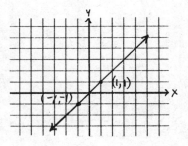

10.

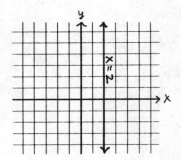

<u>FORM A</u> (continued)

11. Function
    Domain: $\{2, 3, 5\}$
    Range: $\{1, 4\}$

12. Not a function

13. Function
    Domain: All reals
    Range: $\{11\}$

14. II

15. I

16. 7, −1

17. $\frac{2}{3}, -\frac{5}{3}$

18. 3

19. 7

20. $\frac{2}{3}$

ANSWERS TO CHAPTER 7 TESTS (continued)

<u>FORM B</u>

1.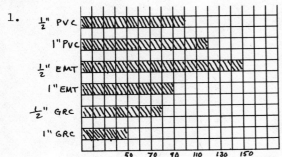

2.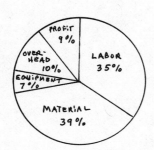

3.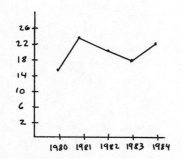

4. right triangle

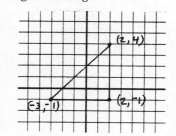

5. square

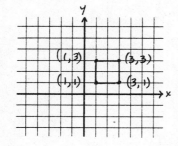

6.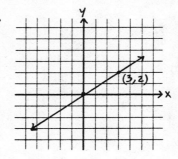

7.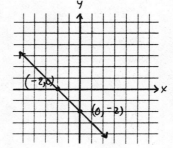

8.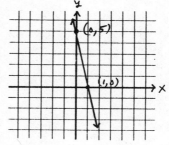

9.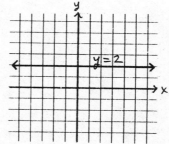

10.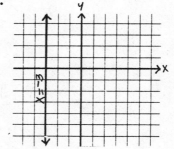

ANSWERS TO CHAPTER 7 TESTS (continued)

<u>FORM B</u> (continued)

11. Function
    Domain: $\{5, 6\}$
    Range: $\{5, 6\}$

12. Not a function

13. Not a function

14. III

15. II

16. −2, 11

17. $-\frac{4}{9}$, $-\frac{1}{3}$

18. −3

19. 0

20. $-\frac{2}{7}$

FORM C

1.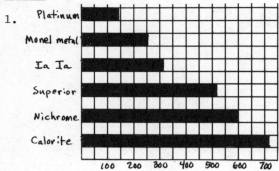

2.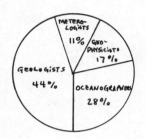

3.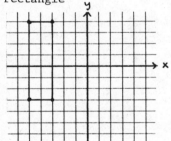

4. rectangle

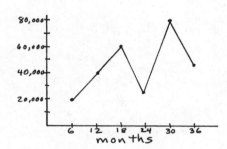

5. triangle

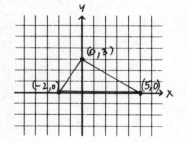

6.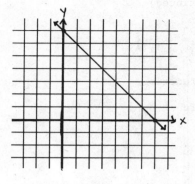

7.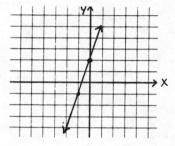

8.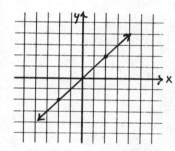

9.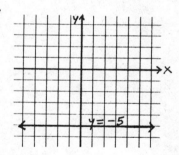

10.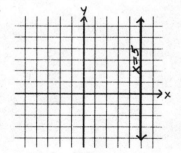

116

11. Function
    Domain: $\{1, 3, 4\}$
    Range: $\{2\}$

12. Not a function
    Domain: $\{1, 2, 5\}$
    Range: $\{1, 2, 3, 6\}$

13. Function
    Domain is the set of reals
    Range: $\{6\}$

14. I

15. IV

16. $-8, 5$

17. $-\frac{1}{5}, 1$

18. -3

19. 0

20. $-\frac{3}{7}$

FORM D

1.

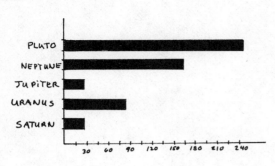

2.

3.

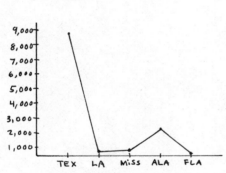

4.

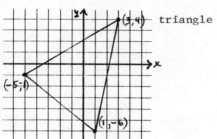 triangle

5.

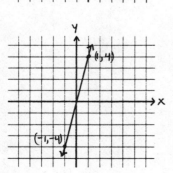

6.

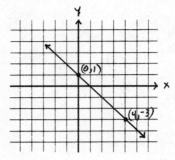

7.

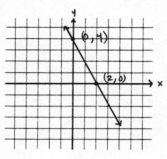

8.

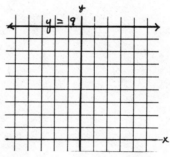

9. Function
    Domain: $\{6, 8\}$
    Range: $\{8\}$

10. Function
    Domain: $\{-3, -1, 2, 7\}$
    Range: $\{1, 2, 3, 7\}$

11. Not a function
    Domain: $\{5\}$
    Range is set of real numbers

12.

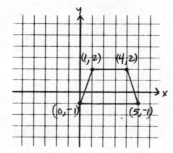

FORM D (continued)

13. $12m^2$

14. II

15. I

16. 11, −11

17. 2, −20

18. −16

19. $\frac{1}{3}$

20. no slope

CHAPTER 8 PRETEST

8.1   In problems 1 through 4, solve the following systems by graphing:

1. $x + y = 1$
      $x + 2y = 0$

2. $a - b = 1$
       $2a - 3b = 0$

3. $x + 2y = 4$
     $-2x + y = -8$

4. $2D + E = -2$
       $D - 2E = 4$

8.2   In problems 5 through 8, solve using the addition method:

5. $A + B = 0$
     $2A - B = -6$

6. $3x + y = -4$
      $4x + 3y = 8$

7. $2X - 3Y = 17$
     $5X - 7Y = 49$

8. $4e - 6f = 3$
     $2e + 4f = 1$

In problems 9 through 12, solve by substitution:

9. $X + 5Y = -11$
     $2X - Y = 11$

10. $3M - N = 8$
      $M + 5N = -8$

11. $3T - 2V = -16$
      $3T + 7V = 29$

12. $4a - 5b = -9$
      $7a - 5b = -12$

8.3   In problems 13 through 16, evaluate:

13. $\begin{vmatrix} 2 & 4 \\ 8 & -1 \end{vmatrix}$

14. $\begin{vmatrix} 5 & -1 \\ -9 & 3 \end{vmatrix}$

15. $\begin{vmatrix} 5 & -1 \\ 3 & -8 \end{vmatrix}$

16. $\begin{vmatrix} 6 & 10 \\ 3 & 5 \end{vmatrix}$

In problems 17 through 20, solve by Cramer's Rule:

17. $3j - 5k = 7$
     $7j - 11k = 13$

18. $-4c - 3d = 17$
      $9c + 22d = -4$

19. $-19X + 23Y = 16$
       $5X + 7Y = -10$

20. $15X - 16Y = 20$
      $-2X + 5Y = -10$

8.4   In problems 21 and 22, evaluate:

21. $\begin{vmatrix} 2 & -1 & 3 \\ 5 & 4 & -2 \\ 8 & -1 & 1 \end{vmatrix}$

22. $\begin{vmatrix} 0 & 4 & 7 \\ -2 & 6 & 0 \\ 8 & 0 & -3 \end{vmatrix}$

In problems 23 through 25, solve by Cramer's Rule:

23. $2x + 3y - z = -15$
     $-4x + 5y + z = -7$
     $3x - y - z = -4$

24. $-e + f - g = -1$
     $8e - 2f + 5g = 5$
     $5e - 3f + 4g = 4$

25. $2X + 2Y + 2Z = 2$
     $3X + 5Y - Z = -11$
     $-X - Y + 4Z = 9$

CHAPTER 8 TEST
FORM A

Solve the following linear systems and tell whether they are independent, inconsistent, or dependent. If consistent, state the solution.

1. Solve by graphing:

$x + 2y = 3$
$3x - y = -5$

1._____

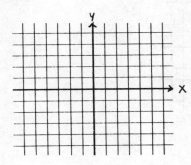

2. Solve by the addition method:

$3x - 2y = -7$
$4x + 5y = -17$

2._____

3. Solve by substitution:

$4R + 3T = 10$
$2R - T = -10$

3._____

4. Solve by using Cramer's Rule:

$x - 2y = 5$
$3y + y = -2$

4._____

5. Solve by using Cramer's Rule:

$3x - y + z = 10$
$2x + y + z = 6$
$x + 3y + 2z = 5$

5._____

Solve problems 6 through 10 using any method:

6. $.6a - .2b = .06$
$.5a + .1b = .13$

6._____

7. $3/5x - 2/3y = 1/6$
$-18x + 20y = -1$

7._____

8. $2I_1 - 5I_2 - 20 = 0$

$3I_1 - I_2 + 9 = 0$

8._____

9. $7x - 5y = -6$
$3x - 3y = -4$

9._____

10. Six gallons of Grade A-1 paint are mixed with four gallons of Grade A-2 paint to obtain a mixture worth $56.92. Three gallons of the Grade A-1 paint are mixed with five gallons of the Grade A-2 paint to obtain a mixture worth $46.40. Find the price of a gallon of Grade A-1 paint and the price of a gallon of Grade A-2 paint.

10._____

Solve the following linear systems, and tell whether they are independent, inconsistent, or dependent. If consistent, state the solution.

1. Solve by graphing:

   $y - 4 = 0$
   $4x - y = 0$

   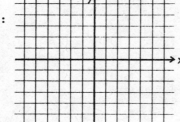

   1._____

2. Solve by the addition method:

   $5x - 3y = -4$
   $2x - 4y = 4$

   2._____

3. Solve by substitution:

   $3I_1 - I_2 = 15$

   $5I_1 + 2I_2 = 14$

   3._____

4. Solve by using Cramer's Rule:

   $6x + 8y = 3$
   $-4x + 10y = -25$

   4._____

5. Solve by using Cramer's Rule:

   $2x - 3y - z = 11$
   $x + y + z = 0$
   $3x - y - 2z = -6$

   5._____

In problems 6 through 10, solve using any method:

6. $\frac{2}{3} R_1 - \frac{5}{8} R_2 = 1$ 7. $0.4x + 0.3y = 0.09$
                                                $0.5x - 0.2y = 0.17$

   $16R_1 - 15R_2 = 3$

   6._____

   7._____

8. $\frac{3}{2} A - 3B = \frac{15}{2}$

   $-\frac{5}{2} A + 5B = -\frac{25}{2}$

   8._____

9. The sum of two numbers is 140. Two times the smaller number is 20 less than the larger number. Find the numbers.

   9._____

10. The perimeter of a rectangle is 48 cm. If four times the width of the rectangle is subtracted from three times the length of the rectangle, the difference is 30 cm. Find the length and width of the rectangle.

    10._____

Solve the following linear systems, and tell whether they are
independent, inconsistent, or dependent. If consistent, state
the solution.

1. Solve by graphing:

   $x + 3y = -3$
   $2x + y = 4$

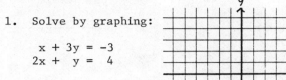

1._____

2. Solve by the addition or subtraction method:

   $a + b = 6$
   $a + 3b = 20$

2._____

3. Solve by using the substitution method:

   $4R - T = 13$
   $3R - 2T = 16$

3._____

4. Solve by using Cramer's Rule:

   $6e - 9f = 29$
   $3e + 6f = -10$

4._____

5. Solve by using Cramer's Rule:

   $4x + y - 2z = 2$
   $3x - 5y + z = 14$
   $5x + 2y - z = 10$

5._____

In problems 6 through 10, solve using any method:

6. $\dfrac{x + 6}{15} - \dfrac{y + 2}{10} = 1$

   $\dfrac{x + 2}{4} - \dfrac{y - 4}{6} = \dfrac{10}{3}$

7. $\dfrac{3}{4} x - \dfrac{2}{3} y = \dfrac{1}{6}$

   $9x - 8y = 2$

6._____

7._____

8. $5c - 2d = -7$
   $2c + 3d = 1$

9. $\dfrac{1}{3} P - \dfrac{1}{5} Q = \dfrac{5}{6}$
   $10P - 6Q = 1$

8._____

9._____

10. The sum of two numbers is 18. One number is 6 more than one-
    fifth of the other. Find the numbers.

10._____

Solve the following linear systems, and tell whether they are
independent, inconsistent, or dependent. If consistent, state
the solution.

1. Solve by graphing:

    x + 4y = 7
    x + 5y = 8

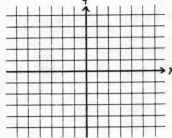

1._____

2. Solve by the addition or subtraction method:

    2H − 3K = 11
    5H + 7K = −16

2._____

3. Solve by using the substitution method:

    −x + 2y = −4
    2x + y = −2

3._____

4. Solve by using Cramer's Rule:

    −2x + 3y = 5
     7x − 5y = 11

4._____

5. Solve by using Cramer's Rule:

    3x + 2y − 4z = −6
     x − 5y − 8z = 14
    −2x + 3y + 6z = −10

5._____

For problems 6 through 10, solve using any method:

6. 2D − 3E = −4
    6D − 9E = 6

7. $\dfrac{2}{5} x - \dfrac{4}{5} y = \dfrac{6}{25}$

    $\dfrac{-3}{7} x + \dfrac{6}{7} y = -\dfrac{9}{35}$

8. $\dfrac{1}{6} e - \dfrac{5}{3} f = -\dfrac{17}{2}$

    $\dfrac{2}{3} e - \dfrac{1}{2} f = -\dfrac{19}{6}$

9. $\dfrac{C + 3}{9} - \dfrac{D + 5}{2} = 0$

    $\dfrac{C + 6}{12} + \dfrac{2D - 10}{8} = -1$

6._____

7._____

8._____

9._____

10. The sum of two numbers is 40, and one number is 5 less than
    twice the other number. Find the numbers.

10._____

ANSWERS TO CHAPTER 8 TESTS

<u>PRETEST</u>

1. (2, −1)

2. (3, 2)

3. (4, 0)

4. (0, −2)

5. (−2, 2)

6. (−4, 8)

7. (28, 13)

8. $(\frac{3}{2}, -\frac{1}{2})$

9. (4, −3)

10. (2, −2)

11. (−2, 5)

12. (−1, 1)

13. −34

14. 6

15. −37

16. 0

17. (−6, −5)

18. $(-\frac{362}{61}, \frac{137}{61})$

19. $(-\frac{171}{124}, -\frac{55}{124})$

20. $(-\frac{60}{43}, -\frac{110}{43})$

21. −86

22. −360

23. (−1, −3, 4)

24. $(\frac{1}{5}, -\frac{1}{5}, \frac{3}{5})$

25. (2, −3, 2)

<u>FORM A</u>

1. (−1, 2)

2. (−3, −1)

3. (−2, 6)

4. (1/7, −17/7)

5. (2, −1, 3)

6. (.2, .3)

7. Inconsistent

8. (−5, −6)

9. (1/3, 5/3)

10. Grade A−1 is $5.50
    a gallon

    Grade A−2 is $5.98
    a gallon

<u>FORM B</u>

1. (1, 4)

2. (−2, −2)

3. (4, −3)

4. (5/2, −3/2)

5. (−1, −1, 2)

6. Inconsistent

7. (0.3, −0.1)

8. Dependent

9. 40 and 100

10. 18 cm by 6 cm

ANSWERS TO CHAPTER 8 TESTS (continued)

FORM C

1. (3, −2)

2. (−1, 7)

3. (2, −5)

4. ($\frac{4}{3}$, −$\frac{7}{3}$)

5. (3, 0, 5)

6. (6, −4)

7. Dependent

8. (−1, 1)

9. Inconsistent

10. 8 and 10

FORM D

1. (3, 1)

2. (1, −3)

3. (0, −2)

4. ($\frac{58}{11}$, $\frac{57}{11}$)

5. (2, −4, 1)

6. Inconsistent

7. Dependent

8. (−1, 5)

9. (6, −3)

10. 15 and 25

9.1 Simplify:

1. $(x^4y^2)(x^5y^3) =$ 2. $\dfrac{x^{20}}{x^{13}} =$ 3. $\dfrac{(a^3b^2)^4}{(a^2b)^3}$

9.2 Simplify:

4. $\left(-\dfrac{3}{5}\right)^0$ 5. $x^4 \cdot x^0 \cdot x^3 =$ 6. $\dfrac{1}{(x-y)^0}$

9.3 Simplify and give answers without negative exponents:

7. $\dfrac{1}{5^{-2}} =$ 8. $4m^{-2}$

9. $\dfrac{x^{-2}}{x^{-3}} =$ 10. $(2ab^2)^{-2}$

9.4 Simplify and express with positive exponents only:

11. $a^{1/5}a^{2/5} =$ 12. $\dfrac{x^{6/7}}{x^{2/7}} =$

13. $(x^{1/3})^{1/3}$ 14. $\dfrac{a^{5/8}b^{3/4}}{a^{1/8}b^{1/4}} =$

9.5 Express the following numbers in scientific notation:

15. 5,200,000   16. 0.00054

Express the following as ordinary numbers:

17. 3×10^5 18. 2.5×10^{-4}

9.6 Perform the following calculations using scientific notation:

19. $(30,000)(0.005)$ 20. $\dfrac{60,000}{0.0003}$

I. Simplify:

1. $c \cdot c^4 \cdot c^3$

2. $\dfrac{e^8}{e^5}$

3. $(m^3)^6$

4. $(2a^3b)^2(3^2ab^3)^3$

5. $\dfrac{(2c^3d)^2}{(4cd^3)^2}$

6. $7a^0$

7. $-3(x^0)$

8. $\dfrac{m^{-2}}{n^{-3}}$

9. $(-3R^{-1}S^2T^{-3})^{-2}$

10. $\dfrac{(5R^2)^{-3}}{(2R^{-3})^3}$

11. $T^{3/4} \cdot T^{1/6}$

12. $\dfrac{f^{2/3}}{f^{1/5}}$

13. $(H^{1/7})^{\frac{1}{2}}$

14. $\dfrac{M^{-1/2}N^{-2/3}}{M^{-3/2}N}$

15. $(2x^{-2}y)^{-2}(-3xy^3)^{-1}$

16. $a^2 - b^{-1}$

1._____

2._____

3._____

4._____

5._____

6._____

7._____

8._____

9._____

10._____

11._____

12._____

13._____

14._____

15._____

16._____

II. Perform the following calculations using scientific notation. Express answers in scientific notation.

17. Round 0.4863 to three significant digits and express it in scientific notation.

18. $(0.00014)(24,000)$

19. $\dfrac{0.0012}{2,400,000}$

20. $\dfrac{(16,000,000)(0.360)(0.030)}{(1,800,000)(0.000032)(0.0005)}$

17._____

18._____

19._____

20._____

CHAPTER 9 TEST
FORM B

I. Simplify:

1. $x^3 \cdot x^8$

2. $2^3 \cdot 2$

3. $\dfrac{10^{11}}{10^4}$

4. $(p^3)^5$

5. $(a^2b^3)^3(2ab)^4$

6. $\dfrac{(-2xy^2)^3}{2x^2y^8}$

7. $2^3 \cdot (-1)^0$

8. $9D^0$

9. $\dfrac{3}{3^{-2}}$

10. $\dfrac{(2c^2)^{-2}}{(3c^{-1})^{-1}}$

11. $2f^{-2}(2f^2)^{-3}$

12. $\dfrac{-2}{(2A)^{-2}}$

13. $4a^0 - 2e^{-2}$

14. $\dfrac{k^{5/7}}{k^{1/3}}$

15. $\dfrac{R^{-1/2}T^{1/2}}{R^{1/2}T^2}$

16. $x^{1/3} \cdot x^{1/3}$

17. If $x^{-2/3}$ is a factor of $x^{1/4}$, what is the other factor?

18. $(x^{1/4}y^{1/2})^{1/4}$

II. Use scientific notation to calculate:

19. $(16,000)(28,000,000)$

20. $\dfrac{(72,000)(36,000)}{(1,200)(40,000,000)(0.008)}$

1._____

2._____

3._____

4._____

5._____

6._____

7._____

8._____

9._____

10._____

11._____

12._____

13._____

14._____

15._____

16._____

17._____

18._____

19._____

20._____

I.   Simplify:

1. $y^3 \cdot y^7$

2. $\dfrac{5^6}{5^4}$

3. $(5ie^2)^2$

4. $\left(\dfrac{2}{x^2}\right)^3$

5. $(3ab^2)^5(3^2a^4bc)^3$

6. $5H^0$

7. $(-3K)^0$

8. $\dfrac{-2}{k^{-1}}$

9. $(-3B)^{-2}$

10. $6f^{-1}j^2$

11. $\left(\dfrac{2B^3}{3BC^{-1}}\right)^{-2}$

12. $x - y^{-2}$

13. $(-2j^{-2}m)^{-3}(2jm^{-3})^2$

14. $a^{1/2}\,a^{1/2}$

15. $\dfrac{b^{3/5}}{b^{1/2}}$

16. $(cd^2)^{1/2}$

17. $\left(E^{1/2}\right)^{1/4}$

18. $\dfrac{p^{-3/4}Q^{-1/2}}{p^{1/2}Q^{-1/3}}$

19. $(x^{-1/2}y^{1/2})^{1/2}$

20. Calculate using scientific notation and express the answer in scientific notation:

$$\dfrac{(25,000,000)(0.16)(0.00003)}{(9,000)(3.2)(0.0005)}$$

1._____

2._____

3._____

4._____

5._____

6._____

7._____

8._____

9._____

10._____

11._____

12._____

13._____

14._____

15._____

16._____

17._____

18._____

19._____

20._____

Simplify the following:

1. $b^2 b^5$

2. $3^2 \cdot 3$

3. $\dfrac{d^3}{d^8}$

4. $(3ir^2)^3$

5. $(4fh^2)^3(4^2f^3h)^4$

6. $\dfrac{(2RT^2)^3}{2RT^2}$

7. $(10xy)^0$

8. $8e^0$

9. $\dfrac{1}{a^{-3}}$

10. $-3j^{-2}$

11. $\dfrac{-1}{(3R)^{-2}}$

12. $\dfrac{(6m^3)^{-2}}{(2m^{-2})^{-1}}$

13. $a - b^{-2}$

14. $(5t^2)^{-1}(25t^{-3})^{-1}$

15. $d^{3/4} \cdot d^{1/4}$

16. $\dfrac{R^{1/2}}{R^{1/4}}$

17. $(3E^{-1/2}H^{1/4})^{-1/2}$

18. The volume of a sphere is $V = \dfrac{\pi d}{6}$, where d is the diameter. Find the volume of a sphere whose radius is $2r^3$.

19. Round 0.007851 to two significant figures and express it in scientific notation.

20. Calculate using scientific notation, and express the answer in scientific notation:

$$\dfrac{(0.00015)(0.000002)}{(0.08)(450,000)}$$

1. _____

2. _____

3. _____

4. _____

5. _____

6. _____

7. _____

8. _____

9. _____

10. _____

11. _____

12. _____

13. _____

14. _____

15. _____

16. _____

17. _____

18. _____

19. _____

20. _____

ANSWERS TO CHAPTER 9 TESTS

PRETEST

1. $x^9 y^5$
2. x^7
3. $a^6 b^5$
4. 1
5. x^7
6. 1
7. 25
8. $\dfrac{4}{m^2}$
9. x
10. $\dfrac{1}{4a^2 b^4}$
11. $a^{3/5}$
12. $x^{5/7}$
13. $x^{1/9}$
14. $a^{1/2} b^{1/2}$
15. 5.2×10^6
16. 5.4×10^{-4}
17. 300,000
18. 0.00025
19. $1.5 \times 10^2 = 150$
20. $2 \times 10^8 = 200,000,000$

FORM A

1. c^8
2. e^3
3. m^{18}
4. $2916\, a^9 b^{11}$
5. $\dfrac{c^4}{4d^4}$
6. 7
7. -3
8. $\dfrac{n^3}{m^2}$
9. $\dfrac{R^2 T^6}{9S^4}$
10. $\dfrac{R^3}{1000}$
11. $T^{11/12}$
12. $f^{7/15}$
13. $H^{1/14}$
14. $\dfrac{M}{N^{5/3}}$
15. $\dfrac{-x^3}{12y^5}$
16. $\dfrac{a^2 b - 1}{b}$
17. 4.86×10^{-1}
18. 3.36
19. 5×10^{-10}
20. 6×10^6

FORM B

1. x^{11}
2. 16
3. 10^7
4. p^{15}
5. $16a^{10} b^{13}$
6. $\dfrac{-4x}{y^2}$
7. 8
8. 9
9. 27
10. $\dfrac{3}{4c^5}$
11. $\dfrac{1}{4f^8}$
12. $-8A^2$
13. $\dfrac{4e^2 - 2}{e^2}$
14. $k^{8/21}$
15. $\dfrac{1}{RT^{3/2}}$
16. $x^{2/3}$
17. $x^{11/12}$
18. $x^{1/16} y^{1/8}$
19. 4.48×10^{11}
20. 6.8×10^2

ANSWERS TO CHAPTER 9 TESTS (continued)

FORM C

1. y^{10}

2. 25

3. $25i^2e^4$

4. $\dfrac{8}{x^6}$

5. $3^{11}a^{17}b^{13}c^3$

6. 5

7. 1

8. $-2k$

9. $\dfrac{1}{9B^2}$

10. $\dfrac{6j^2}{f}$

11. $\dfrac{9}{4B^4C^2}$

12. $\dfrac{xy^2 - 1}{y^2}$

13. $\dfrac{-j^8}{2m^9}$

14. a

15. $b^{1/10}$

16. $c^{1/2}d$

17. $E^{1/8}$

18. $\dfrac{1}{p^{5/4}Q^{1/6}}$

19. $\dfrac{y^{1/4}}{x^{1/4}}$

20. 8.3×10^{0}

FORM D

1. b^7

2. 27

3. $\dfrac{1}{d^5}$

4. $27i^3r^6$

5. $4^{11}f^{15}h^{10}$

6. $4R^2T^4$

7. 1

8. 8

9. a^3

10. $\dfrac{-3}{j^2}$

11. $-9R^2$

12. $\dfrac{1}{18m^8}$

13. $\dfrac{ab^2 - 1}{b^2}$

14. $\dfrac{t}{125}$

15. d

16. $R^{1/4}$

17. $\dfrac{E^{1/4}}{3^{1/2}H^{1/8}}$

18. $V = \dfrac{2\pi r^3}{3}$

19. 7.9×10^{-3}

20. 8.3×10^{-15}

10.1 Find the following roots:

1. $\sqrt[3]{64}$ 2. $\sqrt{0.64}$ 3. $\sqrt{-16}$

10.2 Find the roots. Assume that all variables represent non-negative numbers:

4. $\sqrt{x^{12}}$ 5. $\sqrt[3]{y^9}$ 6. $\sqrt{16x^8}$

7. $\sqrt{a^8b^{10}c^4}$

8. Express $2^{3/5}$ in radical form.

9. Reduce the index of $\sqrt[8]{y^6}$.

10. Evaluate: $27^{-1/3}$.

10.3 Simplify the following radicals:

11. $\sqrt{80}$ 12. $\sqrt[3]{81}$ 13. $-2\sqrt{24x^5y^3}$

10.4 Multiply the following radicals. Simplify wherever possible.

14. $\sqrt{5} \cdot \sqrt{12}$ 15. $(-2\sqrt{5})(-3\sqrt{6})$ 16. $\sqrt{x^3y} \cdot \sqrt{x^2y^3} \cdot \sqrt{xy^4}$

10.5 Simplify the following:

17. $\dfrac{\sqrt{20}}{\sqrt{5}}$ 18. $\dfrac{\sqrt[3]{x}}{\sqrt[3]{x^2}}$

10.6 19. $3\sqrt{18} + 2\sqrt{32} =$ 20. $5\sqrt{a^3} + 6\sqrt{a^5} =$

10.7 Multiply and simplify answers wherever possible:

21. $(2 + \sqrt{3})(5 - \sqrt{3})$ 22. $(4 - \sqrt{2})^2$

10.8 Rationalize:

23. $\dfrac{\sqrt{5}}{\sqrt{3} + 1}$ 24. $\dfrac{\sqrt{3} - 1}{\sqrt{3} + 1}$

CHAPTER 10 TEST
FORM A

1. Find the principal root of $\sqrt[4]{16p^4}$.

2. Express $\sqrt[3]{a}$ in exponential form.

3. Express $R^{3/7}$ in radical form.

4. Reduce the index of $\sqrt[8]{s^2}$.

In problems 5 through 7, simplify:

5. $-3\sqrt{108}$

6. $\sqrt[3]{m^3 n^4}$

7. $-3\sqrt[5]{-32}$

In problems 8 through 20, perform the indicated operations and simplify if possible:

8. $\sqrt{2} \cdot \sqrt{6}$

9. $\sqrt[5]{d^4} \cdot \sqrt[5]{d^3}$

10. $\sqrt{6}(\sqrt{3} - 2\sqrt{5})$

11. $(3\sqrt{2} + \sqrt{5})(2\sqrt{2} - 3\sqrt{5})$

12. $\dfrac{2t}{\sqrt[3]{t^2}}$

13. $\dfrac{\sqrt{7}}{(\sqrt{3} + 4)}$

14. $\sqrt{11} - 2\sqrt{11} + 6\sqrt{11}$

15. $\dfrac{2}{3}\sqrt{\dfrac{2}{3}} - \dfrac{1}{9}\sqrt{24}$

16. $5\sqrt{y} - 2\sqrt{y}$

17. $2\sqrt{45} - 4\sqrt{5} + \sqrt{20}$

18. $\dfrac{3}{\sqrt{6}}$

19. $\dfrac{\sqrt{3} - 2\sqrt{2}}{\sqrt{3} + 2\sqrt{2}}$

20. If $c = \sqrt{a^2 + b^2}$, find c if a = 4 and b = 6.

1. _____

2. _____

3. _____

4. _____

5. _____

6. _____

7. _____

8. _____

9. _____

10. _____

11. _____

12. _____

13. _____

14. _____

15. _____

16. _____

17. _____

18. _____

19. _____

20. _____

135

1. Express $a^{3/4}$ in radical form.

     1._____

2. Express $\sqrt[7]{c^3}$ in exponential form.

     2._____

3. Give the principal square root of 100.

     3._____

In problems 4 through 6, simplify:

4. $2\sqrt{90a^3b^4c}$

     4._____

5. $-3\sqrt{108}$

     5._____

6. $-3\sqrt[5]{-32}$

     6._____

In problems 7 through 20, perform the indicated operations and simplify if possible:

7. $\sqrt{3} \cdot \sqrt{5}$

     7._____

8. $-6\sqrt[3]{27}$

     8._____

9. $\dfrac{\sqrt{2r}}{\sqrt{3r}}$

     9._____

10. $\sqrt{2}(\sqrt{5} - \sqrt{2})$

     10._____

11. $(3\sqrt{a} - 2\sqrt{b})(3\sqrt{a} + 5\sqrt{b})$

     11._____

12. $(\sqrt{2} - \sqrt{3})^2$

     12._____

13. $5\sqrt{3} + \sqrt{3} - 3\sqrt{3}$

     13._____

14. $2\sqrt[3]{16b^4} - 3b\sqrt[3]{2b}$

     14._____

15. $\dfrac{\sqrt{3}}{\sqrt{7}}$

     15._____

16. $\dfrac{\sqrt[3]{y}}{\sqrt[3]{x^2}}$

     16._____

17. $\dfrac{1}{3}\sqrt{\dfrac{2}{3}} - \dfrac{1}{9}\sqrt{24} + \sqrt{\dfrac{3}{2}}$

     17._____

18. $6\sqrt{45} - 10\sqrt{20} - 2\sqrt{80}$

19. $\dfrac{\sqrt{3}}{(\sqrt{2} + 7)}$

     19._____

20. $\dfrac{(2\sqrt{5} - 3\sqrt{2})}{(2\sqrt{5} + 3\sqrt{2})}$

     20._____

CHAPTER 10 TEST
FORM C

1. Find the principal square root of 144.

2. Simplify: $\sqrt[3]{8t^{12}}$

3. Express $\sqrt[5]{d^3}$ in exponential form.

4. Express $a^{2/3}$ in radical form.

5. Reduce the index of $\sqrt[4]{x^2}$

In problems 6 and 7, simplify the following:

6. $-\frac{3}{2}\sqrt{128}$

7. $\sqrt[3]{16a^4b^2c^7}$

In problems 8 through 20, perform the indicated operations.
Simplify the answers when possible.

8. $\sqrt{6y} \cdot \sqrt{2y}$

9. $\sqrt[3]{3} \cdot \sqrt[3]{7}$

10. $\dfrac{\sqrt[3]{18}}{\sqrt[3]{2}}$

11. $\sqrt{5}(\sqrt{3} - \sqrt{2})$

12. $\dfrac{\sqrt{c}}{\sqrt{d}}$

13. $\dfrac{\sqrt{5}}{\sqrt{2}}$

14. $5\sqrt{3} - 2\sqrt{3}$

15. $2\sqrt{2} + 3\sqrt{\frac{1}{2}} - 4\sqrt{\frac{1}{4}}$

16. $(3\sqrt{5} + \sqrt{6})(4\sqrt{5} - 2\sqrt{6})$

17. $(3 - \sqrt{D})^2$

18. $\dfrac{1}{\sqrt{3} - 1}$

19. $\dfrac{2\sqrt{6} + \sqrt{2}}{2\sqrt{6} - \sqrt{2}}$

20. $h\sqrt{\frac{1}{h}} - 5\sqrt{h}$

1._____

2._____

3._____

4._____

5._____

6._____

7._____

8._____

9._____

10._____

11._____

12._____

13._____

14._____

15._____

16._____

17._____

18._____

19._____

20._____

1. Express in radical form: $x^{1/4}$

1._____

2. Give the principal square root of 121.

2._____

3. Express $\sqrt[6]{a^5}$ in exponential form.

3._____

In problems 4 through 7, simplify:

4. $\sqrt{175}$

4._____

5. $\sqrt[6]{y^3}$

5._____

6. $\sqrt{f^7}$

6._____

7. $3\sqrt{20x^2y^3z^4}$

7._____

In problems 8 through 20, perform the indicated operations and simplify if possible:

8. $\sqrt{3} \cdot \sqrt{11}$

8._____

9. $\sqrt[3]{9} \cdot \sqrt[3]{3}$

9._____

10. $\sqrt{5}(\sqrt{2} - \sqrt{8})$

10._____

11. $(5\sqrt{3} - \sqrt{2})^2$

11._____

12. $6\sqrt{x} - 2\sqrt{x}$

12._____

13. $2\sqrt{y} + 5\sqrt{y} - 7\sqrt{y}$

13._____

14. $5\sqrt{3} - 2\sqrt{27} + \sqrt{12}$

14._____

15. $\dfrac{2}{\sqrt{5}}$

15._____

16. $\dfrac{7}{\sqrt[3]{x^2}}$

16._____

17. $\dfrac{3}{\sqrt{6} - 2}$

17._____

18. $\dfrac{\sqrt{2} - 3\sqrt{5}}{\sqrt{2} + 3\sqrt{5}}$

18._____

19. $\dfrac{1}{3}\sqrt{\dfrac{2}{3}} - \dfrac{1}{9}\sqrt{24} + \sqrt{\dfrac{3}{2}}$

19._____

20. If $s = \sqrt{\dfrac{r}{t}}$, find s when r = 384 and t = 3.

20._____

ANSWERS TO CHAPTER 10 TESTS

1. 4

2. 0.8

3. not a real number

4. x^6

5. y^3

6. $4x^4$

7. $a^4b^2c^2$

8. $\sqrt[5]{8}$

9. $\sqrt[4]{y^3}$

10. $\dfrac{1}{3}$

11. $4\sqrt{5}$

12. $3\sqrt[3]{3}$

13. $-4x^2y^2\sqrt{6xy}$

14. $2\sqrt{15}$

15. $6\sqrt{30}$

16. x^3y^4

17. 2

18. $\dfrac{3\sqrt{x^2}}{x}$

19. $17\sqrt{2}$

20. $a\sqrt{5a} + a^2\sqrt{6a}$

21. $7 + 3\sqrt{3}$

22. $18 - 8\sqrt{2}$

23. $\dfrac{\sqrt{15} - \sqrt{5}}{2}$

24. $2 - \sqrt{3}$

FORM A

1. $2p$

2. $a^{1/3}$

3. $7\sqrt{R^3}$

4. $4\sqrt{s}$

5. $-18\sqrt{3}$

6. $mn\sqrt[3]{n}$

7. 6

8. $2\sqrt{3}$

9. $d\sqrt[5]{d^2}$

10. $3\sqrt{2} - 2\sqrt{30}$

11. $-7\sqrt{10 - 3}$

12. $2\sqrt[3]{t}$

13. $\dfrac{\sqrt{21} - 4\sqrt{7}}{-13}$

14. $5\sqrt{11}$

15. 0

16. $3\sqrt{y}$

17. $4\sqrt{5}$

18. $\dfrac{\sqrt{6}}{2}$

19. $\dfrac{-11 + 4\sqrt{6}}{5}$

20. $c = 2\sqrt{13}$

FORM B

1. $\sqrt[4]{a^3}$

2. $c^{3/7}$

3. 10

4. $6ab^2 \sqrt{10ac}$

5. $-18 \sqrt{3}$

6. 6

7. $\sqrt{15}$

8. -18

9. $\dfrac{\sqrt{6}}{3}$

10. $\sqrt{10} - 2$

11. $9a + 9\sqrt{ab} - 10b$

12. $5 - 2\sqrt{6}$

13. $3\sqrt{3}$

14. $b\sqrt[3]{2b}$

15. $\dfrac{\sqrt{21}}{7}$

16. $\dfrac{\sqrt[3]{xy}}{x}$

17. $\dfrac{7}{18}\sqrt{6}$

18. $-10\sqrt{5}$

19. $\dfrac{7\sqrt{3} - \sqrt{6}}{47}$

20. $19 - 6\sqrt{10}$

FORM C

1. 12

2. $2t^4$

3. $d^{3/5}$

4. $\sqrt[3]{a^2}$

5. $\sqrt{x}$

6. $-12\sqrt{2}$

7. $2ac^2 \sqrt[3]{2ab^2c}$

8. $2y\sqrt{3}$

9. $\sqrt[3]{21}$

10. $\sqrt[3]{9}$

11. $\sqrt{15} - \sqrt{10}$

12. $\dfrac{\sqrt{cd}}{d}$

13. $\dfrac{\sqrt{10}}{2}$

14. $3\sqrt{3}$

15. $\dfrac{7\sqrt{2}}{2} - 2$

16. $48 - 2\sqrt{30}$

17. $9 - 6\sqrt{D} + D$

18. $\dfrac{\sqrt{3} + 1}{2}$

19. $\dfrac{13 + 4\sqrt{3}}{11}$

20. $-4\sqrt{h}$

FORM D

1. $\sqrt[4]{x}$

2. 11

3. $a^{5/6}$

4. $5\sqrt{7}$

5. $\sqrt{y}$

6. $f\sqrt[3]{f}$

7. $6xyz^2 \sqrt{5y}$

8. $\sqrt{33}$

9. 3

10. $-\sqrt{10}$

11. $77 - 10\sqrt{6}$

12. $4\sqrt{x}$

13. 0

14. $\sqrt{3}$

15. $\dfrac{2\sqrt{5}}{5}$

16. $\dfrac{7\sqrt[3]{x}}{x}$

17. $\dfrac{3\sqrt{6} + 6}{2}$

18. $\dfrac{6\sqrt{10} - 47}{43}$

19. $\dfrac{7\sqrt{6}}{18}$

20. $8\sqrt{2}$

CHAPTER 11 PRETEST

11.1 Solve by taking the square root of both members of the equation:

1. $b^2 = 128$ 2. $(R - 2)^2 = 18$

11.2 Solve by completing the square:

3. $a^2 + a = 3$ 4. $2X^2 + 2X - 7 = 0$

11.3 Solve by using the quadratic formula:

5. $m^2 + 6m - 10 = 0$ 6. $3Y^2 - Y = 10$

7. $\dfrac{x + 1}{x + 2} - \dfrac{2x - 3}{x - 2} = 0$

11.4 Simplify:

8. $\sqrt{-81}$ 9. $\sqrt{-48}$ 10. $3 + 2\sqrt{-49}$

Add or subtract the given complex numbers. Express the answers in the form of a + bj.

11. $(6 - 2j) + (5 - j)$ 12. $(3 - 8j) - (-2 - j)$

11.5 Perform the indicated operations and express the answers in the form of a + bj.

13. $3(5 + 2j)$ 14. $j(5 - 4j)$

15. $3j(-8 - 3j)$ 16. $(2 + 5j)(9 - j)$

17. $\dfrac{3j}{1 + 5j}$ 18. $\dfrac{2 - 7j}{4 + 3j}$

11.6 Solve for the given variable and express answers in the form of a + bj.

19. $x^2 = -24$ 20. $3N^2 = -2N - 10$

Solve the following quadratic equations by taking the square root of both members of the equation:

1. $t^2 - 18 = 0$ 2. $(2p - 7)^2 = 3$

3. $T^2 = -27$

Solve by completing the square:

4. $3A^2 + 2A - 7 = 0$ 5. $6Y^2 - 5Y + 1 = 0$

Solve by the quadratic formula:

6. $2x^2 - 2x - 5 = 0$ 7. $7R^2 - R - 1 = 0$

Solve by using any method:

8. The sum of two voltages is 15, and the sum of their squares is 173. Find the voltages.

9. $35M^2 + 34M = 21$

10. $\dfrac{x + 5}{x - 2} - 3 = \dfrac{x - 1}{x + 2}$

11. $9x^2 - 3x - 11 = 0$

Express the following in complex form:

12. $\sqrt{-121}$ 13. $4 - 2\sqrt{-9}$

Perform the following operations for the given complex numbers:

14. $2(4 - j) + j(3 - 2j)$

15. $(2 - 7j) + (10 + j)$

16. $(3 - j) - (4 - 2j)$

17. $(1 - 5j)(2 + 3j)$

18. $\dfrac{2 + 3j}{3 - j}$

Solve over the set of complex numbers. Express answers in the form of a + bj.

19. $3X^2 + X + 1 = 0$

20. $9Y^2 - 2Y + 11 = 0$

1._____

2._____

3._____

4._____

5._____

6._____

7._____

8._____

9._____

10._____

11._____

12._____

13._____

14._____

15._____

16._____

17._____

18._____

19._____

20._____

CHAPTER 11 TEST
FORM B

Solve the following quadratic equations by taking the square root of both members of the equation:

1. $B^2 = 8$

2. $(T - 3)^2 = 64$

3. $R^2 = -162$

Solve by completing the square:

4. $x^2 + 2x - 7 = 0$

5. $3x^2 - 5X - 1 = 0$

Solve by the quadratic formula:

6. $6T^2 - 5T - 6 = 0$

7. $7R^2 - R - 2 = 0$

Solve by using any method:

8. $\dfrac{x + 2}{x - 1} - 4 = \dfrac{x + 5}{x + 1}$

9. The length of a rectangle is 5 inches more than the width. The area of the rectangle is 85 square inches. Find the dimensions of the rectangle.

10. $15x^2 + 19x = 10$

11. $2d^2 - 5d - 9 = 0$

Express the following in complex form:

12. $\sqrt{-49}$

13. $6 - 3\sqrt{-25}$

Perform the following operations for the given complex numbers.

14. $3(5 - j) + j(4 - 3j)$

15. $(4 - 5j) - (6 - j)$

16. $(11 - 14j) + (13 + 5j)$

17. $(3 - j)(2 + j)$

18. $\dfrac{5 + 2j}{5 - 2j}$

Solve over the set of complex numbers. Express answers in the form of a + bj.

19. $2X^2 - 3X + 5 = 0$

20. $2e^2 - 3e = -10$

1. _____

2. _____

3. _____

4. _____

5. _____

6. _____

7. _____

8. _____

9. _____

10. _____

11. _____

12. _____

13. _____

14. _____

15. _____

16. _____

17. _____

18. _____

19. _____

20. _____

Solve the following quadratic equations by taking the square root of both members of the equation:

1. $3A^2 = 4$

2. $(5T - 2)^2 = 18$

3. $Y^2 = -44$

Solve by completing the square:

4. $x^2 + 2x - 9 = 0$

5. $6b^2 + b = 1$

Solve by the quadratic formula:

6. $2E^2 - 6E + 1 = 0$

7. $3y^2 - 3y - 2 = 0$

Solve using any method:

8. $\dfrac{x + 3}{x + 1} - 5 = \dfrac{x + 3}{x + 2}$

9. A garden covers 220 square meters. It is 12 meters longer than it is wide. Find the length and the width of the garden.

10. $4X^2 - 3X = 9$

11. $2A^2 - 3A - 12 = 0$

Express the following in complex form:

12. $\sqrt{-48}$

13. $3 - 2\sqrt{-144}$

Perform the following operations for the given complex numbers:

14. $(2 + 5j) + (3 - 4j)$

15. $(2 + 5j) - (3 - 4j)$

16. $4(2 - 2j) + j(5 - j)$

17. $(2 + 5j)(3 - 4j)$

18. $\dfrac{2 + 5j}{3 - 4j}$

Solve over the set of complex numbers. Express answers in the form of a + bj.

19. $X^2 - 2X + 5 = 0$

20. $5Y^2 - 5Y = -9$

1._____

2._____

3._____

4._____

5._____

6._____

7._____

8._____

9._____

10._____

11._____

12._____

13._____

14._____

15._____

16._____

17._____

18._____

19._____

20._____

Solve the following quadratic equations by taking the square root of both members of the equation:

1. $2d^2 = 9$ 2. $(M - 4)^2 = 3$

3. $h^2 = -45$

1._____

2._____

3._____

Solve by completing the square:

4. $Y^2 - 3Y - 1 = 0$ 5. $7X^2 - 2X - 3 = 0$

4._____

5._____

Solve by using the quadratic formula:

6. $x^2 - 5x - 2 = 0$ 7. $9E^2 - 2E - 3 = 0$

6._____

7._____

Solve the following quadratic equations using any method:

8. $\dfrac{x + 1}{x - 2} - 3 = \dfrac{x + 4}{x + 1}$

8._____

9. $3B^2 - B = 6$

9._____

10. $6C^2 = -7C + 3$

10._____

11. The altitude of a triangle is 3 meters less than twice the base. Find the altitude if the area is 297 square meters.

11._____

Express the following in complex form:

12. $\sqrt{-18}$ 13. $4 - 5\sqrt{-16}$

12._____

13._____

Perform the following operations for the given complex numbers:

14. $7(2 - 3j) + 2j(5 - j)$ 15. $(3 + 2j) + (7 - j)$

16. $(3 + 2j) - (7 - j)$ 17. $(3 + 2j)(7 - j)$

18. $\dfrac{3 + 2j}{7 - j}$

14._____

15._____

16._____

17._____

18._____

Solve over the set of complex numbers:

19. $X^2 - X + 7 = 0$

20. $3y^2 - 2y + 5 = 0$

19._____

20._____

ANSWERS TO CHAPTER 11 TESTS

1. $b = \pm 8\sqrt{2}$

2. $R = 2 \pm 3\sqrt{2}$

3. $a = \dfrac{-1 \pm \sqrt{13}}{2}$

4. $X = \dfrac{-1 \pm \sqrt{15}}{2}$

5. $m = -3 \pm \sqrt{19}$

6. $Y = 2$ or $Y = -\dfrac{5}{3}$

7. $x = -1 \pm \sqrt{5}$

8. $9j$

9. $4j\sqrt{3}$

10. $3 + 14j$

11. $11 - 3j$

12. $5 - 7j$

13. $15 + 6j$

14. $4 + 5j$

15. $9 - 24j$

16. $23 + 43j$

17. $\dfrac{15}{26} + \dfrac{3}{26}j$

18. $-\dfrac{13}{25} - \dfrac{34}{25}j$

19. $X = \pm 2j\sqrt{6}$

20. $N = -\dfrac{1}{3} \pm \dfrac{\sqrt{29}}{3}j$

FORM A

1. $t = \pm 3\sqrt{2}$

2. $p = \dfrac{7 \pm \sqrt{3}}{2}$

3. $T = \pm 3j\sqrt{3}$

4. $A = \dfrac{-1 \pm \sqrt{22}}{3}$

5. $Y = \dfrac{1}{2}$, $Y = \dfrac{1}{3}$

6. $x = \dfrac{1 \pm \sqrt{11}}{2}$

7. $R = \dfrac{1 \pm \sqrt{29}}{14}$

8. 2 and 13

9. $M = \dfrac{3}{7}$ or $M = -\dfrac{7}{5}$

10. $x = \dfrac{5 \pm \sqrt{85}}{3}$

11. $x = \dfrac{1 \pm 3\sqrt{5}}{6}$

12. $11j$

13. $4 - 6j$

14. $10 + j$

15. $12 - 6j$

16. $-1 + j$

17. $17 - 7j$

18. $\dfrac{3}{10} + \dfrac{11}{10}j$

19. $X = -\dfrac{1}{6} \pm \dfrac{\sqrt{11}}{6}j$

20. $Y = \dfrac{1}{9} \pm \dfrac{7\sqrt{2}}{9}j$

ANSWERS TO CHAPTER 11 TESTS (continued)

FORM B

1. $B = \pm 2\sqrt{2}$

2. $T = 11$ or $T = -5$

3. $R = \pm 9j\sqrt{2}$

4. $x = -1 \pm 2\sqrt{2}$

5. $X = \dfrac{5 \pm \sqrt{37}}{6}$

6. $T = -\dfrac{2}{3}$ or $T = \dfrac{3}{2}$

7. $R = \dfrac{1 \pm \sqrt{57}}{14}$

8. $x = \dfrac{1 \pm \sqrt{177}}{8}$

9. 7 in. by 12 in.

10. $x = \dfrac{2}{5}$ or $x = -\dfrac{5}{3}$

11. $d = \dfrac{5 \pm \sqrt{97}}{4}$

12. $7j$

13. $6 - 15j$

14. $18 + j$

15. $-2 - 4j$

16. $24 - 9j$

17. $7 + j$

18. $\dfrac{21}{29} + \dfrac{20}{29}j$

19. $X = \dfrac{3}{4} \pm \dfrac{\sqrt{31}}{4}J$

20. $e = \dfrac{3}{4} \pm \dfrac{\sqrt{71}}{4}j$

FORM C

1. $A = \dfrac{\pm 2\sqrt{3}}{3}$

2. $T = \dfrac{2 \pm 3\sqrt{2}}{5}$

3. $Y = \pm 2j\sqrt{11}$

4. $x = -1 \pm \sqrt{10}$

5. $b = \dfrac{1}{3}$ or $b = -\dfrac{1}{2}$

6. $E = \dfrac{3 \pm \sqrt{7}}{2}$

7. $y = \dfrac{3 \pm \sqrt{33}}{6}$

8. $x = \dfrac{-7 \pm \sqrt{14}}{5}$

9. 22m by 10m

10. $X = \dfrac{3 \pm \sqrt{153}}{8}$

11. $A = \dfrac{3 \pm \sqrt{105}}{4}$

12. $4j\sqrt{3}$

13. $3 - 24j$

14. $5 + j$

15. $-1 + 9j$

16. $9 - 3j$

17. $26 + 7j$

18. $-\dfrac{14}{25} + \dfrac{23}{25}j$

19. $X = 1 \pm 2j$

20. $Y = \dfrac{1}{2} \pm \dfrac{\sqrt{155}}{10}$

FORM D

1. $d = \dfrac{\pm\, 3\,\sqrt{2}}{2}$

2. $M = 4 \pm \sqrt{3}$

3. $h = \pm\, 3j\,\sqrt{5}$

4. $Y = \dfrac{3 \pm \sqrt{13}}{2}$

5. $X = \dfrac{1 \pm \sqrt{22}}{7}$

6. $x = \dfrac{5 \pm \sqrt{33}}{2}$

7. $E = \dfrac{1 \pm\, 2\,\sqrt{7}}{9}$

8. $x = \dfrac{1 \pm \sqrt{21}}{2}$

9. $B = \dfrac{1 \pm \sqrt{73}}{6}$

10. $C = -\dfrac{3}{2}$ or $C = \dfrac{1}{3}$

11. $33m$

12. $3j\,\sqrt{2}$

13. $4 - 20j$

14. $16 - 11j$

15. $10 + j$

16. $-4 + 3j$

17. $23 + 11j$

18. $\dfrac{19}{50} + \dfrac{17}{50}\,j$

19. $X = \dfrac{1}{2} \pm \dfrac{3j\,\sqrt{3}}{2}$

20. $y = \dfrac{1}{3} \pm j\sqrt{\dfrac{14}{3}}$

CHAPTER 12 PRETEST

12.1   1. Graph the following exponential function: $y = 3^x$.

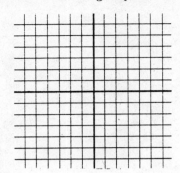

2. Graph the following logarithmic function: $x = \left(\dfrac{1}{2}\right)^y$.

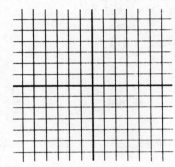

12.2   3. Express $2^{-5} = \dfrac{1}{32}$ in logarithmic form.

4. Express $\log_2 \dfrac{1}{16} = -4$ in exponential form.

5. Solve: $\log_2 \dfrac{1}{4} = x$

12.3   6. Find log 38.5.

7. Find log 0.0634.

12.4   8. Find the antilog of 3.7210.

9. Find the antilog of 0.9058 − 1.

12.5 10. Use the properties of logarithms to find $\log(302)(21.6)$.

11. Use the properties of logarithms to find $\log(7)^2$.

12. Solve for x: $\log 2 + \log x = \log (x + 1)$.

12.6 Compute the following using logarithms:

13. $(36.8)(15,000)$

14. $\sqrt{106}$

12.7 Find ln 257.

For problems 1 and 2, graph:

1. $y = \left(\frac{1}{2}\right)^x$

2. $x = (4)^y$

For problems 3 and 4, express in exponential form:

3. $\log 7 \, \frac{1}{49} = -2$

4. $\log_5 125 = 3$

For problems 5 and 6, express in logarithmic form:

5. $7^3 = 343$

6. $10^{-1} = \frac{1}{10}$

For problems 7 through 9, use the definition of logs to find the logarithms of:

7. $\log_3 1 = x$

8. $\log_2 \frac{1}{32} = x$

9. $\log_8 512 = x$

For problems 10 through 12, use Table 2 to find the common logarithm approximation of the following:

10. $\log 86.3$

11. $\log 0.00084$

12. $\log 986{,}000{,}000$

For problems 13 through 16, use Table 2 to find the anti-logarithmic approximation of the following:

13. antilog of 1.7284

14. antilog of 0.0755–2

15. antilog of 1.6314

16. Find: ln 39

In problems 17 and 18, use the properties of logarithms to find:

17. $\log 69 \sqrt{8}$

18. $\log \dfrac{\sqrt{7}}{6.8}$

In problems 19 through 22, use logarithms to compute the following:

19. $\dfrac{29.5}{0.00486}$

20. $\sqrt[4]{82.1}$

21. $(0.0721)^6$

22. $(4{,}200)(875)$

In problems 23 through 25, solve for x:

23. $2^{3x-1} = 5$

24. $\log (3x - 4) + \log 2 = \log 4$

25. $\log (x + 5) - \log (x - 3) = \log 3$

1._____

2._____

3._____

4._____

5._____

6._____

7._____

8._____

9._____

10._____

11._____

12._____

13._____

14._____

15._____

16._____

17._____

18._____

19._____

20._____

21._____

22._____

23._____

24._____

25._____

CHAPTER 12 TEST
FORM B

For problems 1 and 2, graph:

1. $y = 5^x$

2. $x = \left(\frac{1}{3}\right)^y$

For problems 3 and 4, express in logarithmic form:

3. $14^2 = 196$

4. $64^{-1/3} = \frac{1}{4}$

For problems 5 and 6, express in exponential form:

5. $\log_6 216 = 3$

6. $\log_{16} \frac{1}{2} = -\frac{1}{4}$

For problems 7 through 9, use the definition of a logarithm to find:

7. $\log_{100} 10 = x$

8. $\log_5 \frac{1}{25} = x$

9. $\log_2 32 = x$

For problems 10 through 12, use Table 2 to find the common logarithm approximations. If a calculator is used, give answers approximate to 4 decimal places.

10. $\log 8,370$

11. $\log 78.4$

12. $\log 0.0082$

For problems 13 through 16, use Table 2 to find the anti-logarithmic approximation of the following. If using a calculator, approximate answers to 3 significant digits.

13. antilog of 0.7513

14. antilog of 3.5453

15. antilog of 0.8215–4

16. Find: ln 52

17. Use the properties of logarithms to find: $\log 68 \sqrt[3]{16}$

For problems 18 through 21, use logarithms to compute the following:

18. $(68.4)(32.1)$

19. $\dfrac{0.418}{0.0068}$

20. $(34.1)^3$

21. $\sqrt{695}$

For problems 22 through 25, solve for x:

22. $4^x = 1$

23. $3^{2x + 1} = 6$

24. $2 \log(x - 7) = \log 9$

25. $\log (x + 4) - \log (x - 2) = \log 5$

1._____
2._____
3._____
4._____
5._____
6._____
7._____
8._____
9._____
10._____
11._____
12._____
13._____
14._____
15._____
16._____
17._____
18._____
19._____
20._____
21._____
22._____
23._____
24._____
25._____

151

1. Graph: $y = \log_4 x$

For problems 2 and 3, express in logarithmic form:

2. $3^4 = 81$

3. $8^{-3} = \frac{1}{512}$

For problems 4 and 5, express in exponential form:

4. $\log_5 125 = 3$

5. $\log_{49} \frac{1}{7} = -\frac{1}{2}$

For problems 6 through 8, use the definition of logarithms to find the following:

6. $\log_2 16 = x$

7. $\log_{11} \frac{1}{121} = x$

8. $\log_{27} 3 = x$

For problems 9 through 11, use Table 2 to find the common logarithm approximation of the following. If a calculator is used, approximate to 4 decimal places.

9. log 4.73

10. log 3500

11. log 0.0706

For problems 12 through 15, use Table 2 to find the anti-logarithmic approximation of the following. If a calculator is used, approximate to 3 significant digits.

12. antilog of 0.8007

13. antilog of 3.9143

14. antilog of 0.4425-2

15. Find the ln of 49.

For problems 16 and 17, find the following logs, using the properties of logarithms:

16. $\log 622 \sqrt[3]{9}$

17. $\log \frac{\sqrt{84}}{7.28}$

For problems 18 through 21, use logarithms to compute the following:

18. $(0.0684)(4{,}920)$

19. $\frac{471}{0.28}$

20. $(0.0149)^4$

21. $\sqrt{0.74}$

For problems 22 through 25, solve for x:

22. $2^x = 32$

23. $3^{x+1} = 8$

24. $\log x + \log 5 = \log 20$

25. $\log (x + 1) - \log (x - 6) = \log 8$

1._____
2._____
3._____
4._____
5._____
6._____
7._____
8._____
9._____
10._____
11._____
12._____
13._____
14._____
15._____
16._____
17._____
18._____
19._____
20._____
21._____
22._____
23._____
24._____
25._____

For problems 1 and 2, graph:

1. $y = 3^x$

2. $x = \left(\frac{1}{4}\right)^y$

For problems 3 and 4, express in logarithmic form:

3. $9^2 = 81$

4. $4^{-1/2} = \frac{1}{2}$

For problems 5 and 6, express in exponential form:

5. $\log_3 81 = 4$

6. $\log_5 \frac{1}{25} = -2$

For problems 7 through 9, use the definition of a logarithm to find the following:

7. $\log_{25} 5 = x$

8. $\log_3 \frac{1}{27} = x$

9. $\log_{13} 169 = x$

For problems 10 through 12, use Table 2 to find the common logarithm approximation of the following. If a calculator is used, approximate answers to 4 decimal places.

10. log 924

11. log 68.1

12. log 0.0723

For problems 13 through 16, use Table 2 to find the antilogarithmic approximation of the following. If a calculator is used, approximate answers to 3 significant digits:

13. antilog of 0.7024

14. antilog of 2.6212

15. antilog of .8506-3

16. Find: ln 28

For problems 17 and 18, find the following logs using the properties of logarithms:

17. $\log 78 \sqrt{7}$

18. $\log \frac{\sqrt{8}}{4.2}$

For problems 19 through 22, use logarithms to compute the following:

19. $(386,000)(0.00441)$

20. $\frac{3030}{4690}$

21. $(0.084)^3$

22. $\sqrt[3]{0.752}$

For problems 23 through 25, solve for x:

23. $2 \log(x - 3) = \log 1$

24. $2^{x + 1} = 7$

25. $\log(x + 1) + \log(x - 5) = \log(x^2 - 14)$

1._____

2._____

3._____

4._____

5._____

6._____

7._____

8._____

9._____

10._____

11._____

12._____

13._____

14._____

15._____

16._____

17._____

18._____

19._____

20._____

21._____

22._____

23._____

24._____

25._____

ANSWERS TO CHAPTER 12 TESTS

PRETEST

1.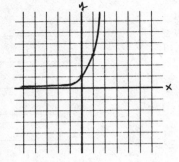

2.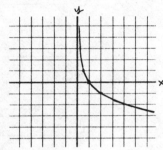

3. $\log_2 \frac{1}{32} = -5$

4. $2^{-4} = \frac{1}{16}$

5. $x = -2$

6. 1.5855

7. 0.8021 − 2

8. 3.7210

9. 0.9058 − 1

10. 3.8145

11. 1.6902

12. $x = 1$

13. 552,000

14. 10.3

15. 5.55

FORM A

1.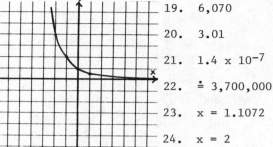

2.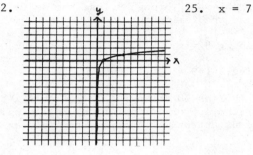

19. 6,070

20. 3.01

21. 1.4×10^{-7}

22. $\doteq 3,700,000$

23. $x = 1.1072$

24. $x = 2$

25. $x = 7$

3. $7^{-2} = \frac{1}{49}$

4. $5^3 = 125$

5. $\log_7 343 = 3$

6. $\log \frac{1}{10} = -1$

7. $x = 0$

8. $x = -5$

9. $x = 3$

10. 1.9360

11. 0.9243 − 4

12. 8.9939

13. 53.5

14. 0.019

15. 42.8

16. 3.664

17. 2.290

18. 0.5901−1

ANSWERS TO CHAPTER 12 TESTS (continued)

FORM B

1.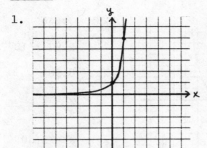

2.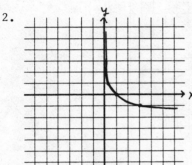

18. 2,200

19. 61.5

20. 39,700

21. 26.4

22. x = 0

23. 0.3155

24. x = 10 or x = 4

25. x = 7/2

3. $\log_{14} 196 = 2$

4. $\log_{64} \frac{1}{4} = -\frac{1}{3}$

5. $6^3 = 216$

6. $16^{-1/4} = \frac{1}{2}$

7. $x = \frac{1}{2}$

8. x = -2

9. x = 5

10. 3.9227

11. 1.8943

12. 0.9138-3

13. 5.64

14. 3,510

15. 0.000663

16. 3.951

17. 2.2339

FORM C

1.
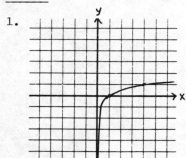

2. $\log_3 81 = 4$

3. $\log_8 \frac{1}{512} = -3$

4. $5^3 = 125$

5. $49^{-1/2} = 1/7$

6. $x = 4$

7. $x = -2$

8. $x = 1/3$

9. 0.6749

10. 3.5441

11. 0.8488−2

12. 6.32

13. 8,210

14. 0.0277

15. 3.892

16. 3.1119

17. 0.1001

18. 337

19. 1700

20. 4.9×10^{-8}

21. 0.860

22. $x = 5$

23. $x = 0.8928$

24. $x = 4$

25. $x = 7$

FORM D

1.
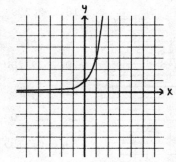

2.
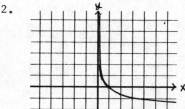

3. $\log_9 81 = 2$

4. $\log_4 \frac{1}{2} = -\frac{1}{2}$

5. $3^4 = 81$

6. $5^{-2} = \frac{1}{25}$

7. $x = \frac{1}{2}$

8. $x = -3$

9. $x = 2$

10. 2.9657

11. 1.8331

12. 0.8591−2

13. 5.04

14. 418

15. 0.00709

16. 3.332

17. 2.3147

18. 0.8284−1

19. 1,700

20. 0.646

21. 5.93×10^{-4}

22. 0.909

23. $x = 2$ or
    $x = 4$

24. $x = 1.8074$

25. $x = \frac{9}{4}$

13.1   1. Give the quadrant of 184° and give one positive and one negative
            coterminal angle.

        2. Give the complement of 37 ° 10'.

        3. Convert 270 ° to radian measure.

        4. Convert $\frac{3\pi}{5}$ to degree measure.

13.2   5. Given:

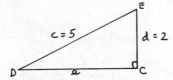

            State the six trigonometric ratios of E.

        6. Given a right triangle, ABC, C = 90°, and csc B = /7, find cos B.

13.3   7. Give cos 47°.

        8. Determine the value of x between 0° and 90° if cot x = 1.343.

13.4   9. Solve the right triangle ABC, given C = 90°, a = 29, and c = 37.

        10. Solve the right triangle ABC, given C = 90°, A = 31°, and c = 52.

13.5   11. Determine the value of cos 169°.

        12. Find θ where 0° ≤ θ < 360°, given cos θ = −0.2079, and tan θ > 0.

13.6   13. (3.0, 84°) is the polar form of a vector. Change the polar form
            to rectangular form correct to three decimal places.

        14. Change the vector (14, −8) from rectangular form to polar form.

13.7   15. Graph the sum of (5 − j) and (2 + 2j).

        16. Graph the vector (2 − j) and the product of j and (2 − j).

        17. Multiply 5 (cos 40° + j sin 40°) and 3(cos 110° + j sin 110°).

13.8   18. Solve triangle ABC using the Law of Sinces, given A = 110°, C = 38°,
            and c = 90.

13.9   19. Solve the triangle ABC, given: a = 3.8, b = 2.5, and c = 1.5.

13.10   20. Graph one period of y = 2 sin $\frac{1}{2}$ x.

        21. Graph one period of y = $\frac{1}{2}$ cos 2x.

        22. Graph one period of y = 3 tan x.

13.11   23. Sketch one cycle of y = 2 cos (x − π).

13.12   24. Solve 2 sin x + 1 = 0 for 0° ≤ x < 360°.

        25. Solve cos θ tan θ + $\frac{1}{2}$ tan θ = 0 for 0 ≤ θ < 2π.

CHAPTER 13 TEST
FORM A

1. Given: 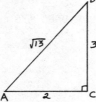 Find: tan A   1._____
                                      sec B

2. Given: csc B = $\frac{\sqrt{41}}{5}$ tan B < 0, find: the exact value   2._____
   of sin A.

3. Use Table 3 or a calculator to determine an approximation   3a._____
   for:
                                                                 b._____
   a) sin 236°20'   b) cot 313°50'   c) cos 301°10'
                                                                 c._____
4. Given: csc x = 1.111, find the angle x between 0° and 90°.   4._____

5. Given: sin x = -.5616, cos x < 0. Find x where   5._____
   0 ≤ x < 360°.

6. Solve the right triangle ABC, given: C = 90°, a = 10, and b = 24.   6._____

7. Solve the right triangle ABC, given: C = 90°, A = 38°, and b = 500.   7._____

8. From the top of a building, the angle of depression to a point 50   8._____
   feet from the foot of the building is 31°. Find the height of
   the building.

9. Solve the oblique triangle ABC, given: A = 64°30', B = 110°40',   9._____
   and b = 246.

10. Solve the oblique triangle ABC, given: C = 39°30', b = 7.9, and   10._____
    a = 6.3.

    For problems 11 and 12, sketch the graphs of the given curves: 11._____

11. y = 3 tan x   12. y = cos (x - π/3)   12._____

13. Express the vector (2.6, 73°20') in rectangular form.   13._____

14. Express the vector (4.8, 3.6) in polar form.   14._____

15. Add (2 - 4j) + (3 - 2j) graphically.   15._____

16. Multiply 4.2(cos 38° + j sin 38° by 2.1(cos 22° + j sin 22°).   16._____

17. Multiply (3 - 2j) by j^2 and graph each vector.   17._____

    In problems 18 through 20, solve the equations for:
    0° ≤ x < 360°   or   0 ≤ θ < 2π

18. $2 \sin^2\theta + \sin \theta - 1 = 0$ 18._____

19. $4 \cos^2 \theta - 6 \cos \theta = 0$ 19._____

20. $\tan^2 x + 3 \tan x - 2 = 0$ 20._____

1. Given: 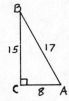   Find: cos A   1._____
                                           csc B

2. Given: cot B = 5/12, sin B $>$ 0, find sec B.   2._____

3. Use Table 3 or a calculator to determine the given   3a._____
    trigonometric ratios:

     b._____

    a)   tan 224°30'   b)   sec 116°20'

     c._____

    c)   cos 325°50'

4. Given: tan x = .4592, find the angle x between 0° and 90°.   4._____

5. Given: cot x = 1.664 and cos x $<$ 0, find x where   5._____
    0° $\leqslant$ x $<$ 360°.

6. Solve the right triangle ABC, given: C = 90°, a = 3.06,   6._____
    and b = 5.13.

7. Solve the right triangle ABC, given C = 90°, a = 2.09,   7._____
    and A = 26°50'.

8. A plane flying at 18,000 feet spots a hangar on the ground.   8._____
    If the angle of depression from the plane to the hangar is
    67°10', how far is the hangar from a point directly below
    the plane?

9. Solve the oblique triangle ABC, given: A = 47°, a = 24.4,   9._____
    and b = 32.2.

    For problems 10 and 11, sketch the graphs of the given curves: 10._____

10. 3 cos 1/2x   11.   sin $(x + \frac{\pi}{6})$ 11._____

12. Express the vector (4.2, 116°) in rectangular form.   12._____

13. Express the vector (3.1, −5.6) in polar form.   13._____

14. Solve the oblique triangle ABC, given: a = 24, b = 38, and   14._____
    c = 32.

15. Add (2 + 3j) + (5 − j) graphically.   15._____

16. Divide 46(cos 172° + j sin 172°) by 22(cos 85° + j sin 85°),   16._____
    and express the answer in rectangular form.

17. Graph 3 + 7j and the product of (3 + 7j) and j.   17._____

    In problems 18 through 20, solve the equations for 0° $\leqslant$ x $<$ 360°   18._____
    or 0 $\leqslant$ θ $<$ 2π.

     19._____

18. $\cot^2 \theta + \dfrac{1}{\sqrt{3}} \cot \theta = 0$ 19. $\tan^2 x - 5 \tan x + 6 = 0$ 20._____

20. $12 \sin^2 x + 25 \sin x + 12 = 0$

1. Given triangle ABC, C = 90°, a = 5, and b = 13.
   Find: sin A
          tan B

   1._____

2. Given: tan A = $\frac{15}{8}$, sec A < 0, determine csc A.

   2._____

3. Use Table 3 to determine the indicated trigonometric
   ratios. (If using a calculator, round to 4 decimal
   places.)

   a) cos 184°   b) cot 314°   c) sec 85°

   3a._____
   b._____
   c._____

4. Given cos x = .7153, find the angle x, between 0° and 90°.

   4._____

5. Solve the right triangle ACB, given: ∠C = 90°, a = 2.09,
   and b = 4.11.

   5._____

6. Solve the right triangle ACB, given: ∠C = 90°, ∠A = 42°,
   and a = 3.0.

   6._____

7. A plane flying at 5400 meters spots a hangar on the ground.
   If the angle of depression from the plane to the hangar is
   67°10', how far is the hangar from a point directly below
   the plane?

   7._____

8. Solve the oblique triangle ABC, given: a = 3.6, b = 6.8,
   and c = 5.2.

   8._____

9. Find θ where 0° ≤ θ < 360°, given tan θ = 1.385, cos θ < 0.

   9._____

   In problems 10 and 11, sketch the graphs of the given curves:

   10._____

10. y = 2 sin $\frac{1}{3}$ x   11. y = cos(x − $\frac{\pi}{3}$)

   11._____

12. Express the vector (2.6, 73°20') in rectangular form.

   12._____

13. Solve the oblique triangle ABC, given: A = 64°30', B = 110°40',
    and b = 246.

   13._____

14. Add (4 − j) and (2 + 4j) graphically.

   14._____

15. Graph: (−2 + 3j) and the product of j and (−2 + 3j).

   15._____

16. Multiply 1.6(cos 120° + j sin 120°) by 2.3(cos 94° + j sin 94°),
    and express the answer in rectangular form.

   16._____

17. Divide 4.22(cos 312° + j sin 312°) by 2.78(cos 54° + j sin 54°),
    and express the answer in rectangular form.

   17._____
   18._____

   In problems 18 through 20, solve the following equations for
   0° ≤ x < 360°, or for 0 ≤ θ < 2π.

   19._____

18. $\cot^2 \theta + \frac{1}{\sqrt{3}}$ cot θ = 0

   20._____

19. $\tan^2 x$ − 5 tan x + 6 = 0

20. 12 $\sin^2 x$ + 25 sin x + 12 = 0

160

CHAPTER 13 TEST
FORM D

1. Given: triangle ABC, C = 90°, a = 15, and b = 8.
   Find: sin A
          csc B

2. Given: $\tan B = \dfrac{12}{5}$, csc B < 0, determine csc A.

3. Use Table 3 to determine the indicated trigonometric
   ratios. (If using a calculator, round to 4 decimal places.)

   a) sin 318°10'   b) cos 420°   c) csc 244°

4. Given: tan x = .6494, find the angle x between 0° and 90°.

5. Solve the right triangle ACB, given: C = 90°, a = 3.06,
   and b = 5.13.

6. Solve the right triangle ACB, given: C = 90°, A = 50°,
   and c = 26.

7. From the top of a building, the angle of depression to a
   point 15.25 meters from the foot of the building is 31°.
   Find the height of the building.

8. Find θ where 0° ≤ θ < 360°, given sin θ = .7071 and
   cos θ < 0.

9. Solve the oblique triangle, given: a = 20, b = 30, and
   c = 40.

10. Sketch the graphs of the given curves:

    a) $-4 \sin \dfrac{1}{2} x$ b) $y = \cos(3x - \dfrac{3\pi}{2})$

11. Express the vector (5.4, 184°) in rectangular form.

12. Express the vector (-2.2, 6.1) in polar form.

13. Solve the oblique triangle ABC, given: A = 47°, a = 24.4,
    and b = 32.2.

14. Add (5 + 3j) and (3 - 2j) graphically.

15. Graph: (3 - 5j) and the product of j and (3 - 5j).

16. Multiply 2.1(cos 62° + j sin 62°) by 3.3(cos 44° + j sin 44°).

17. Divide 6.4 (cos 124° + j sin 124°) by 2.8(cos 36° + j sin 36°).

    In problems 18 through 20, solve the equations for 0° ≤ x < 360°,
    or for 0 ≤ θ < 2π.

18. $\csc^2 \theta = 4$

19. $2 \sin^2 \theta + \sin \theta - 1 = 0$

20. $\tan^2 x + 2 \tan x - 5 = 0$

1._____
2._____
3a._____
 b._____
 c._____
4._____
5._____
6._____
7._____
8._____
9._____
10._____
11._____
12._____
13._____
14._____
15._____
16._____
17._____
18._____
19._____
20._____

PRETEST

1. III, 544°, −176°

2. 52°50'

3. $\dfrac{3\pi}{2}$

4. 108°

5. $\dfrac{\sqrt{21}}{5}$, $\dfrac{2}{5}$, $\dfrac{\sqrt{21}}{2}$, $\dfrac{5\sqrt{21}}{21}$, $\dfrac{5}{2}$, $\dfrac{2\sqrt{21}}{21}$

6. $\dfrac{\sqrt{42}}{7}$

7. 0.6820

8. 36°40'

9. A = 51°40', B = 38°20', b = 23

10. B = 59°, a = 27, b = 45

11. −0.9816

12. 258°

13. (0.314, 2.98)

14. (16, 330°20')

15.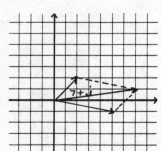

16.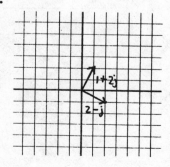

17. 15(cos 150° + j sin 150°)

18. B = 32°, a = 137, b = 77

19. A = 142°, B = 24°, C = 14°

20.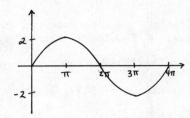

21.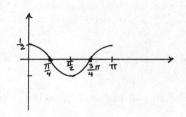

22.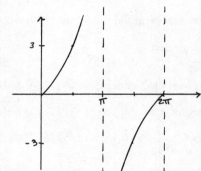

23.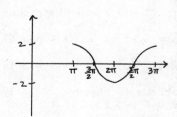

24. 210°, 330°

25. 0, $\dfrac{2\pi}{3}$, $\dfrac{4\pi}{3}$

ANSWERS TO CHAPTER 13 TESTS (continued)

<u>FORM A</u>

1.   3/2, $\dfrac{\sqrt{13}}{3}$

2. $\dfrac{4}{\sqrt{41}}$

3a. -.8323

 b. -.9601

 c. .5175

4.   64°10'

5.   214°10'

6.   B = 67°20'
     A = 22°40'
     c = 26

7.   B = 52°
     a = 394
     C = 634

8.   30 ft

9.   C = 4°50'
     a = 237
     c = 22.2

10. c = 5.0
     A = 52°50'
     B = 87°40'

11.

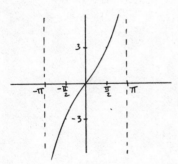

12.

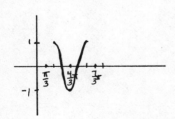

13. (.75, 2.5)

14. (6, 36°50')

15.

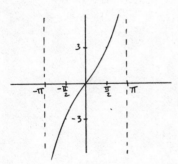

16. 8.8(cos 60° + j sin 60°)

17.

18. θ = $\dfrac{\pi}{6}$, $\dfrac{5\pi}{6}$, $\dfrac{3\pi}{2}$

19. θ = $\dfrac{\pi}{2}$, $\dfrac{3\pi}{2}$

20. x = 105°40', 285°40',
     29°20', 209°20'

FORM B

1. $\frac{8}{17}$, $\frac{17}{8}$

2. $\frac{13}{5}$

3a. .9827

 b. -2.254

 c. .8274

4. 24°40'

5. 211°

6. A = 31°
   B = 59°
   c = 5.97

7. b = 4.11
   B = 63°10'
   c = 4.61

8. 7,578 ft

9. B = 75°   B' = 105°
   C = 58°   or   C' = 28°
   c = 28.3   c = 15.7

10.

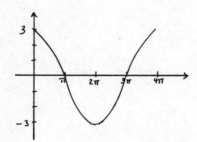

11.

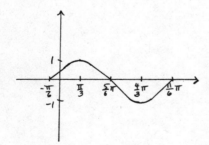

12. (-1.8, 3.8)

13. (6.4, 299°)

14. B = 84°
    A = 39°
    C = 57°

15.

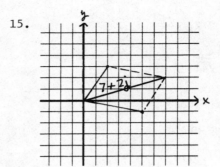

16. (-1.8, 1.1)

17.

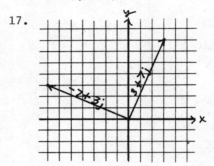

18. $\theta = \frac{\pi}{2}$, $\frac{3\pi}{2}$, $\frac{2\pi}{3}$, $\frac{5\pi}{3}$

19. x = 63°20', 243°20',
       71°30', 251°30'

20. x = 228°40', 311°20'

ANSWERS TO CHAPTER 13 TESTS (continued)

FORM C

1. $\dfrac{5\sqrt{194}}{194}$, $\dfrac{13}{5}$

2. $-17/15$

3a. $-.9976$

  b. $-.9657$

  c. 11.47

4. $44°20'$

5. $A = 26°50'$
   $B = 63°10'$
   $c = 4.61$

6. $B = 48°$
   $c = 4.5$
   $b = 3.3$

7. $2,300$ m

8. $A = 31°$
   $B = 100°$
   $c = 49°$

9. $234°10'$

10.

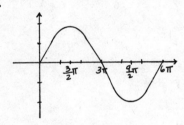

11.

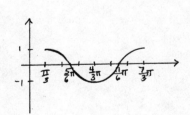

12. $(.75, 2.5)$

13. $C = 4°50'$
   $a = 237$
   $c = 22.2$

14.

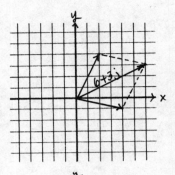

15.

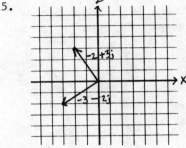

16. $(-3.1, -2.1)$

17. $(-1.49, -.316)$

18. $\dfrac{\pi}{2}$, $\dfrac{3\pi}{2}$, $\dfrac{2\pi}{3}$, $\dfrac{5\pi}{3}$

19. $x = 63°20'$, $243°20'$,
     $71°30'$, $251°30'$

20. $x = 228°40'$, $311°20'$

165

FORM D

1. 15/17, 17/8

2. −13/5

3a. −.6670

 b. .5

 c. −1.113

4. 33°

5. A = 31°
    B = 59°
    c = 5.97

6. B = 40°
    a = 20
    b = 17

7. 9.16 m

8. 135°

9. A = 29°
    B = 47°
    C = 104°

10a.

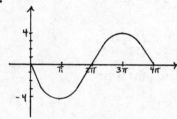

10b.

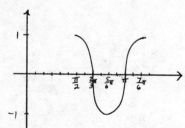

11. (−5.4, .38)

12. (6.5, 109°50')

13. B = 75°   B' = 105°
     C = 58°   or   C' = 28°
     c = 28.3   c' = 15.7

14.

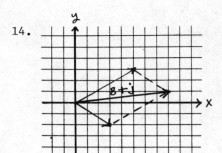

15.

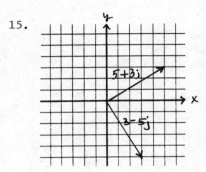

16. (−1.9, 6.7)

17. (1.9, 1.4)

18. $\frac{\pi}{6}$, $\frac{5\pi}{6}$, $\frac{7\pi}{6}$, $\frac{11\pi}{6}$

19. $\frac{\pi}{6}$, $\frac{5\pi}{6}$, $\frac{3\pi}{2}$

20. 55°20', 235°20',
     106°10', 286°10'

166

1.1   1. $296 + 751 + 427 =$ 1._____

2.   Use the commutative property for addition for whole numbers   2._____
to verify that $38 + 27 = 27 + 38$.

3. $5215 - 2768 =$ 3._____

4. $(42 - 20) + (29 - 15) =$ 4._____

1.2   5. $\begin{array}{r} 305 \\ \times\ 46 \\ \hline \end{array}$ 5._____

6. 3×12 years 3 months $=$ 6._____

7.   Use the distributive property to check the following   7._____
problem: $(9 \times 13) + (9 \times 15)$.

8. $35 \div 5 - 6 + 3 \times 8 =$ 8._____

1.3   9.   Write $3 \times 3 \times 3$ using exponents.   9._____

10.   Write $2^2 \times 3^3$ without exponents.   10._____

11.   Express 14,265 in expanded notation using powers of ten.   11._____

1.4   12.   List the prime numbers that are greater than 5 and less   12._____
than 20.

13.   Find the prime factorization of 1260.   13._____

1.5   14.   Express $\dfrac{19}{12}$ as a mixed number.   14._____

15.   Express $9\frac{2}{5}$ as an improper fraction.   15._____

1.6   16.   Write the fraction $\dfrac{1}{7}$ as an equivalent fraction with a   16._____
denominator of 21.

17.   Write the fraction $\dfrac{23}{15}$ as an equivalent fraction with a   17._____
denominator of 60.

1.7   18.   Reduce $\dfrac{14}{18}$ to lowest terms using prime factorization.   18._____

1.8   19.   Write the given fractions as equivalent fractions with   19._____
the lowest common denominator: $\dfrac{3}{5}, \quad \dfrac{5}{8}$

20._____

1.9   20. $\dfrac{4}{7} + \dfrac{5}{7}$ 21. $22\frac{1}{8} - 14\frac{5}{6}$

21._____

1.10   22. $\dfrac{7}{8}$ x $\dfrac{4}{7}$ 23. $6\dfrac{2}{3} \div 2\dfrac{3}{4}$ 22._____

23._____

1.11   24. Write three and twenty-one thousandths in decimal   24._____
             notation.

        25. How many significant digits does 1.06 have?   25._____

        26. Round 4.256 to tenths.   26._____

1.12   Perform the indicated operations. Assume the given
        data are approximations.

        27. 2.52 + 0.007 + 13.03 + 0.7   27._____

        28. 9 - 0.0005   28._____

1.13   Perform the indicated operations. Assume the given
        data are approximations.

        29. 0.42 x 0.24   29._____

        30. 0.49 - 0.7   30._____

        31. Express 13,400 in scientific notation.   31._____

1.14   Perform the indicated operations on the calculator.

        32. 2.35(10 + 40)(70 - 30)   32._____

        33. 42^2 33._____

        34. $5^3 + 3^2 \div 3 + 2^4$ x 5 =   34._____

1.15   35. Express 0.72 as a fraction.   35._____

        36. Express 4.8236 as a fraction.   36._____

        37. Express $\dfrac{4}{5}$ as a decimal.   37._____

1.16   38. Express 0.073 as a percent.   38._____

        39. Express $\dfrac{4}{5}$ as a percent.   39._____

        40. Express 41% as a decimal.   40._____

        41. Express 74% as a fraction.   41._____

1.17   42. Change 3.6 liters to kiloliters.   42._____

        43. Change 5.8 square centimeters to square millimeters.   43._____

        44. Change 25 cubic dekameters to cubic meters.   44._____

        45. Change 20 feet to centimeters.   45._____

2.1   Consider the figure below, given 1 // n, to answer questions
        46, 47, and 48.

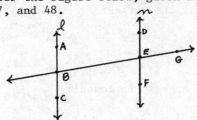

46. Name two rays on line n.   46._____

47. Name a line coincident with $\overleftrightarrow{BE}$.   47._____

48. Name GB in another way.   48._____

2.2   49. Consider the following figure in which 1 // n.   49._____
           Indicate the degree measure of ∠∝ .

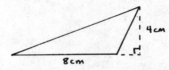

      50. Is an angle whose measure is 91° an acute angle,   50._____
           a right angle, or an obtuse angle?

2.3   51. Find the area of the following triangle: 51._____

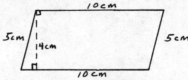

2.4   52. Find the perimeter and area of the following polygon: 52._____

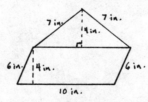

      53. Find the perimeter and area of the following polygon: 53._____

2.5   54. Find the circumference and area of a circle whose diameter 54._____
           is 20 feet.

2.6   55. Find the volume of a right circular cylinder whose base   55._____
           has a diameter of 4.8 cm and whose altitude is 10.6 cm.

3.1   56. Determine which of the given numbers is the largest: 56._____

                9,   −2,   −12

       57. Give the negative of −13.   57._____

       58. Order the following real numbers from highest to lowest: 58._____

                −6, 6, −6 $\frac{1}{4}$, 6.8 −6.7 6.9, −6 $\frac{3}{4}$

3.2   59. Add: 9 + (−7)   59._____

       60. Add: (−1) + (−4) + (−3)   60._____

       61. Find the total of two components (parts) if one component   61._____
            is −23 kg and the other is 25 kg.

3.3   62. Subtract: 10 − 6   62._____

       63. Perform the indicated operations: 3 − 8 + 9 − 2 − 4   63._____

       64. Find the difference between −11 and −31.   64._____

3.4   Add or subtract the following terms:

       65. 103 − 3e   65._____

       66. −9R − 3R   66._____

       67. −8t − (−8t)   67._____

       Add or subtract the following polynomials:

       68. (2F + 2) + (13F − 11)   68._____

       69. (2N − 3M − 16) − (−4M + 2N − 11)   69._____

3.5   70. Multiply: (−9)(−8)   70._____

       71. Divide: (−48) ÷ 6   71._____

3.6   Multiply:

       72. $D^5 \cdot D^2$ 72._____

       73. $2r^6 \cdot 2r^5$ 73._____

       74. $R^3(R + 3)$ 74._____

3.7   75. Divide: $6^9 \div 6^7$ 75._____

3.8   76. 7j(5j + 4) − [2(4j − 2) + 6]   76._____

       77. −[11T + 5(4T − 7)]   77._____

       78. $4x − 3\{2x − 6[x − 2(2x − 5)] + 10\}$ 78._____

3.9   79.   Evaluate $a^2 + 10b - c^2$ if $a = -1$, $b = 2$, and   79._____
            $c = -3$.

      80.   Evaluate $3a - 5b + 4c - 9$ if $a = -1$, $b = 2$,   80._____
            and $c = -3$.

      81.   The formula for the perimeter of a rectangle is   81._____
            $P = 2l + 2w$. Find the perimeter of a rectangle
            with a length of 9 meters and width of 4 meters.

4.1   Solve each equation for the given variable.

      82. $x + 7 = 11$ 82._____

      83. $9e = -18$ 83._____

      84. $2a - 3 = -15$ 84._____

4.2   85. $5(e + 1) = 4e$ 85._____

      86. $0.3T + 0.75 = 1.65$ 86._____

      87. $\dfrac{2M + 1}{4} - \dfrac{5M - 1}{6} = -\dfrac{1}{4}$ 87._____

4.3   Solve the following inequalities and graph the solution
      on the number line.

      88. $20A > 80$ ⟵——————————————⟶   88._____

      89. $\dfrac{3}{7} x < -12$ ⟵——————————————⟶   89._____

      90. $\dfrac{x}{2} - \dfrac{3}{4} < \dfrac{3x}{8}$ ⟵——————————⟶   90._____

4.4   91.   Solve $V = lwh$ for w.   91._____

      92.   Solve $2V_a - V_1 = V_2$ for V_a and find V_a given: 92._____

            $V_1 = 80$ ft/sec, and $V_2 = 24$ ft/sec

      93.   Use the formula from problem 92 to find V_a, given: 93._____

            $V_1 = 96$ ft/sec, and $V_2 = 52$ ft/sec

4.5   94.   Translate the given phrase into an algebraic   94._____
            expression. Let x represent the unknown number.

            The difference of a number and -8.

      In problems 95 and 96, translate each statement into an
      algebraic equation and solve to find the unknown number:

      95.   The sum of a number and 15 is 50.   95._____

      96.   One half of a number subtracted from the number is 3.   96._____

4.6   Solve and check each of the following problems:

97.   The sum of three angles of a triangle equal 180°.   97._____
        If the largest angle is 15° larger than the second
        angle, and the smallest angle is 45° less than the
        degree, measure each angle.

98.   One radio station has a frequency 3.1 times another   98._____
        station. If the difference of the two frequencies
        is 1264.2 kHz, what is the frequency of the higher?

99.   A weight of 180 g is placed 8 cm from the fulcrum of   99._____
        a lever. How far from the fulcrum must a 90 g weight
        be placed for the lever to be balanced?

4.7   100.   Express each of the following as a ratio in lowest   100a._____
                terms:

                a. 16 lb to 24 lb   b. 5 m to 350 cm   b._____

        101.   Find X in the proportion $\frac{11}{16} = \frac{88}{X}$.   101._____

        102.   28 is what percent of 42?   102._____

4.8   103.   Write "R varies inversely as F" as an equation   103._____
                containing a constant of proportionality, k.

        104.   If y varies directly as x, and y = 16 when x = 2,   104._____
                find the proportionality constant.

        105.   If y varies directly as x, and y = 35 when x = 5,   105._____
                find y when x = 20.

        106.   If y varies as x and z, and y = 36 when x = 4 and   106._____
                z = 2, find the proportionality constant.

5.1   Factor the following:

        107.   16x - 24y   107._____

        108. $2M^2 + 5M^4 - 13M^6$ 108._____

        109.   (c + e) - d(c + e)   109._____

5.2   Multiply:

        110.   (x + 6)(x + 2)   110._____

        111.   (j - 9)(j + 9)   111._____

5.3   Factor:

        112. $x^2 + 7x + 6$ 112._____

        113. $a^2b^2 - 9ab + 18$ 113._____

        114. $5b^2 + 4ab - 420$ 114._____

5.4   Factor:

115. $5m^2 + 7m + 2$ 115._____

116. $8A^2 - 10A + 3$ 116._____

117. $20R^2 + 26R - 6$ 117._____

5.5   118. Square the binomial: $(A + 7)^2$ 118._____

119. Factor: $b^2 + 20b + 100$ 119._____

120. Factor: $x^2 - 121$ 120._____

6.1   Reduce to lowest terms:

121. $\dfrac{bx - by}{bx}$ 121._____

122. $\dfrac{2r^2 - 5r + 2}{2r^2 - 7r + 3}$ 122._____

6.2   Perform the indicated operations:

123. $\dfrac{b^2 + b}{b^2 - 4} \cdot \dfrac{b^2 + 5b + 6}{b^2 - 1}$ 123._____

124. $\dfrac{x^2 - y^2}{(x + y)^2} \div \dfrac{x - y}{4x + 4y}$ 124._____

6.3   Find the L.C.D. of the following algebraic fractions and
        convert to equivalent fractions with common denominators:

125. $\dfrac{a}{4x^2} , \dfrac{2a}{6x^3}$ 125._____

126. $\dfrac{2}{b^2 + b - 2} , \dfrac{3}{b^2 - 1}$ 126._____

6.4   Perform the indicated operations and simplify the answers:

127. $\dfrac{5}{x + 3} + \dfrac{2}{x - 2}$ 127._____

128. $\dfrac{c - 1}{c^2 + 3c + 2} - \dfrac{c + 7}{c^2 + 5c + 6}$ 128._____

6.5   Solve and check:

129. $\dfrac{3}{4} + \dfrac{2}{3y} = \dfrac{14}{6y} - \dfrac{1}{12}$ 129._____

130. $\dfrac{4}{y - 1} = \dfrac{7}{y - 5}$ 130._____

7.1   131. The following table gives the maximum output of
              five different wood stoves. The output is measured
              in British thermal units (Ttu) per hour. Construct
              a bar graph to illustrate the maximum output of each
              stove.

                 Woodstove   Btu

                 Hearth   40,000
                 Blazer   20,000
                 Jet II   30,000
                 Archer   20,000
                 Jet III   35,000

                                                                    131._____

7.2   132. Plot the relation $\{(4, -2),(-1, 3), (-2, -2), (2, 3)\}$,
              and identify the geometric figure.

                                                                    132._____

        133. Give the quadrant of each point graphed in problem 132.

                                                                    133._____

7.3   134. Graph the equation $-2x + y = -2$.

                                                                    134._____

7.4   Use the slope-intercept method to graph the following linear
        equations:

        135. $y = -\frac{1}{3}x + 4$

                                                                    135._____

        136. $x - 2y = 10$

                                                                    136._____

8.1   Solve the following systems by graphing, and tell whether they
        are independent, inconsistent, or dependent. If independent,
        give the solution.

        137. $3x + 5y = 2$
              $-3x + y = 4$

                                                                    137._____

        138. $2x - y = -9$
              $2x - y = -3$

                                                                    138._____

8.2   Solve the following systems by using the addition method. Tell
        whether the systems are independent, inconsistent, or dependent.
        if independent, give the solution.

        139. $5X + 3Y = -14$
              $-7X - 3Y = 10$

                                                                    139._____

        140. $5H - 2K = -21$
              $3H - 7K = -1$

                                                                    140._____

        141. Solve the following system by substitution:

              $R + 7T = 5$
              $R - 6T = -8$

                                                                    141._____

8.3   142. Evaluate: $\begin{vmatrix} 7 & 3 \\ 2 & -5 \end{vmatrix}$

                                                                    142._____

        In problems 143 and 144, solve the systems using Cramer's Rule:

        143. $M + N = -1$
                $5M - 10N = 1$

                                                                    143._____

174

144. $2e - 3f = -37$
      $7e + 8f = 0$ 144._____

8.4   145. Evaluate: $\begin{vmatrix} 3 & 4 & 1 \\ -3 & -4 & 1 \\ -2 & -1 & -1 \end{vmatrix}$ 145._____

146. Solve the following system using Cramer's Rule: 146._____

      $3x + 3y + z = 6$
      $3x + 2y - 4z = -22$
      $2x + 5y - z = -12$

9.1   Simplify:

147. $\left(\dfrac{y^4}{y^2}\right)^3$ 147._____

148. $\dfrac{p^8 q^6}{p^4 q^3}$ 148._____

149. $\dfrac{(3x^2 y^4)^3}{(2xy^2)^2}$ 149._____

9.2   Simplify:

150. $\dfrac{2r^0}{3s}$ 150._____

151. $s^2 \cdot s^0 \cdot s^4$ 151._____

9.3   Simplify and give answers without negative exponents:

152. $10^3 \cdot 10^{-5}$ 152._____

153. $\dfrac{x^0 x^4}{x^{-6}}$ 153._____

9.4   Simplify and express answers with positive exponents:

154. $(pq)^{1/3}$ 154._____

155. $\left(\dfrac{A}{B}\right)^{1/2}$ 155._____

9.5   156. Express the following number in scientific notation: 156.

                  0.00009

157. Express 2.3×10^{-5} as an ordinary number.   157._____

158. Express 0.00042 in scientific notation and indicate the number of significant digits.

158._____

9.6 159. Use scientific notation to calculate:

$$\frac{7,500,000,000}{150,000}$$

159._____

10.1 Find the following roots:

160. $\sqrt{36}$

160._____

161. $\sqrt[3]{64}$

161._____

10.2 Assume that all variables represent nonnegative numbers, and find the roots:

162. $\sqrt{B^8}$

162._____

163. $\sqrt[3]{64x^6}$

163._____

164. Express $\sqrt{x^3}$ in exponential form.

164._____

165. Express $B^{2/3}$ in radical form.

165._____

166. Reduce the index of $\sqrt[8]{m^4}$.

166._____

167. Evaluate $32^{-2/5}$.

167._____

10.3 Simplify the following:

168. $\sqrt{45}$

168._____

169. $\sqrt[3]{\frac{4}{27}}$

169._____

170. $4\sqrt[3]{56y^4}$

170._____

10.4 Multiply the following radicals. Simplify wherever possible.

171. $\sqrt{5} \cdot \sqrt{13}$

171._____

172. $2\sqrt{3}(\sqrt{6} - 3\sqrt{5})$

172._____

10.5 Simplify:

173. $\frac{\sqrt{42}}{\sqrt{7}}$

173._____

174. $\frac{5\sqrt{p^3}}{5\sqrt{q^2}}$

174._____

10.6 Add or subtract the following radicals:

175. $4\sqrt{6} + 9\sqrt{6}$

175._____

176. $2\sqrt[3]{9x} + 5\sqrt[3]{9x}$

176._____

177. $\sqrt[3]{x^2y} - \sqrt[3]{8x^2y}$

177._____

10.7   Multiply and simplify answers when possible:

178. $(\sqrt{x} + 3)(\sqrt{x} - 4)$ 178._____

179. $(6\sqrt{2} - 1)(2\sqrt{2} - 3)$ 179._____

10.8   Rationalize:

180. $\dfrac{3}{\sqrt{2} - 1}$ 180._____

181. $\dfrac{\sqrt{A} + \sqrt{B}}{\sqrt{A} - \sqrt{B}}$ 181._____

11.1   Solve by taking the square root of both members of the equation.

182. $R^2 = 25$ 182._____

183. $Y^2 = -25$ 183._____

184. The hypotenuse of a right triangle is 12 and one leg is 2. 184._____
      Find the length of the unknown leg.

11.2   Solve the following by completing the square:

185. $x^2 - 4x - 5 = 0$ 185._____

186. $6R^2 - 5R + 1 = 0$ 186._____

11.3   Solve the following equations using the quadratic formula:

187. $i^2 - 5i - 3 = 0$ 187._____

188. $3x^2 + 7x - 4 = 0$ 188._____

11.4   Simplify:

189. $\sqrt{-144}$ 189._____

In problems 190 and 191, add or subtract the given complex
numbers. Express the answer in the form $a + bj$.

190. $(9 - 2j) + (8 + 5j)$ 190._____

191. $(10 + 2j) + (10 - 2j)$ 191._____

11.5   In problems 192 through 194, perform the indicated operations
        and express answers in the form of $a + bj$.

192. $6(3 - 8j)$ 192._____

193. $5(11 - 4j) + 2j(8 - 3j)$ 193._____

194. $\dfrac{3j}{1 + 2j}$ 194._____

11.6   Solve for the given variable and express answers in the form
        of a + bj.

195. $x^2 = -25$ 195._____

196. $i^2 - 9i + 32 = 0$ 196._____

197. $7y^2 = -3y - 2$ 197._____

12.1   198. Graph the following exponential function: $y = 4^x$ 198._____

        199. Graph the following logarithmic function: $x = 4^y$ 199._____

12.2   200. Express $6^3 = 216$ in logarithmic form.   200._____

        201. Express $8^0 = 1$ in logarithmic form.   201._____

        202. Express $\log_5 5 = 1$ in exponential form. 202._____

        203. Solve for x: $\log_9 81 = x$.   203._____

        204. Solve for y: $\log_{\frac{1}{2}} 16 = y$. 204._____

12.3   Find the logarithm of each of the following:

        205. log 6.5   205._____

        206. log 0.0731   206._____

12.4   207. Find the antilog of 0.8710.   207._____

12.5   Use the properties of logarithms to find the following logs:

        208. $\log (361)(5.28)$ 208._____

        209. $\log (1000)^4$ 209._____

        210. Solve $2^{x-3} = 64$ for x.   210._____

12.6   211. Compute $(4.67)(0.035)$ using logarithms.   211._____

12.7   Find the natural logarithms of the following:

        212. ln 93   212._____

        213. ln 7.53   213._____

13.1   214. State the quadrant of $324°$. Give one positive and one   214._____
              negative coterminal angle for $324°$.

        215. Convert $\frac{5\pi}{8}$ to degree measure. 215._____

13.2   216. Given right triangle ACB with $C = 90°$, and $\cos A = \frac{1}{\sqrt{2}}$,   216._____

              give the other five trigonometric ratios of
              angle A.

178

217. Given right triangle ACB, C = 90°, and sin A = $\frac{1}{2}$, give cot B.

217._____

13.3   218. Use Table 3 or a calculator to find tan 32°20'.

218._____

219. Use Table 3 or a calculator to find the angle x if 0° ≤ x < 90°, and sec x = 2.525.

219._____

13.4   220. Given a right triangle with A = 42°, C = 90°, and c = 9, find angle B and sides a and b.

220._____

13.5   221. Find the exact values of the trigonometric ratios of 240°.

221._____

13.6   222. Change (4.0, 85°) from polar form to rectangular form.

222._____

223. Change (46, 50) from rectangular form to polar form.

223._____

13.7   224. Add 3 + 4j and 6 + 2j by graphing the vectors.

224._____

225. Multiply 2(cos 65° + j sin 65°) by 3(cos 23° + j sin 23°).

225._____

13.8   226. Use the Law of Sines to solve the given triangle.

226._____

A = 125°
B = 25°
a = 30

13.9   227. Solve the oblique triangle, given:

227._____

a = 30
b = 40
c = 60

13.10   228. Graph one period of y = 3 sin $\frac{2}{3}$ x beginning at x = 0.

228._____

13.11   229. Graph one period of y = cos (x + $\frac{\pi}{3}$).

229._____

13.12   230. Solve the equation 2 sin²x − 5 sin x + 2 = 0

230._____

0° ≤ x < 360°.

FINAL EXAMINATION ANSWERS
FORM A

1. 1474

2. 65 = 65

3. 2447

4. 36

5. 14,030

6. 36 years 9 months

7. 9 x (13 + 15) = (9 x 13) + (9 x 15)
   9 x   28   =   117   +   135
              252 = 252

8. 25

9. 3^3

10. 108

11. $(1 \times 10^4) + (4 \times 10^3) + (2 \times 10^2) + (6 \times 10^1) + (5 \times 10^0)$

12. 7, 11, 13, 17, 19

13. $2^2 \times 3^2 \times 5 \times 7$

14. $1 \frac{7}{12}$

15. $\frac{47}{5}$

16. $\frac{3}{21}$

17. $\frac{92}{60}$

18. $\frac{7}{9}$

19. $\frac{24}{40}$, $\frac{25}{40}$

20. $1 \frac{2}{7}$

21. $7 \frac{7}{24}$

22. $\frac{1}{2}$

23. $2 \frac{14}{33}$

24. 3.021

25. three

26. 4.3

27. 16.3

28. 9

29. 0.1

30. 0.7

31. 1.34×10^4

32. 4700

33. 1764

34. 208

35. $\dfrac{18}{25}$

36. $\dfrac{48,236}{10,000}$ or $\dfrac{12059}{2500}$

37. 0.8

38. 7.3%

39. 80%

40. 0.41

41. $\dfrac{37}{50}$

42. 0.0036 kl

43. 580 mm^2

44. 25,000 m^3

45. 609.6 cm

46. $\overrightarrow{ED}$ and $\overrightarrow{EF}$

47. $\overleftrightarrow{BG}$ or $\overleftrightarrow{EG}$

48. $\overrightarrow{GE}$

49. 60°

50. obtuse angle

51. 16 cm^2

52. P = 30 cm, A = 40 cm^2

53. P = 36 in., A = 60 $in.^2$

54. C = 62.8 ft, A = 314 ft^2

55. V $\doteq$ 190 cm^3

56. 9

57. 13

58. 6.9, 6.8, 6, −6, −6 $\frac{1}{4}$, −6.7, −6 $\frac{3}{4}$

59. 2

60. −8

61. 2 kg

62. −3

63. −2

64. 20

65. 7e

66. −12R

67. 0

68. 15F − 9

69. M − 5

70. 72

71. −8

72. D^7

73. $6r^{11}$

74. $R^4 + 3R^3$

75. 6^2

76. $35j^2 + 20j − 2$

77. 35 − 31T

78. −56x + 150

79. 12

80. −34

81. 26m

82. x = 4

83. e = −2

84. a = −6

85. e = −5

86. T = 3

87. M = 2

88. A > 4

89. X < -28

90. X < 6

91. $w = \dfrac{v}{1h}$

92. $V_a = \dfrac{V_1 + V_2}{2} = 52$ ft/sec

93. $V_a = 74$ ft/sec

94. $x - (-8)$

95. $x + 15 = 50$, $x = 35$

96. $x - \dfrac{1}{2} x = 3$

97. $25°$, $70°$, $85°$

98. 1866.2 kHz

99. 16 cm

100. a. $\dfrac{2}{3}$ b. $\dfrac{10}{7}$

101. $x = 28$

102. $66\dfrac{2}{3}$ %

103. $R = \dfrac{k}{F}$

104. $k = 8$

105. $Y = 140$

106. $k = \dfrac{9}{2}$

107. $8(2X - 3Y)$

108. $M^2(2 + 5M^2 - 13M^4)$

109. $(c + e)(1 - d)$

110. $x^2 + 8x + 12$

111. $j^2 - 81$

112. $(x + 6)(x + 1)$

113. $(ab - 2)(ab - 6)$

114. $5(b - 6)(b + 14)$

115. $(5M + 2)(m + 1)$

116. $(4A - 3)(2A - 1)$

117. $2(5R - 1)(2R + 3)$

118. $A^2 + 14A + 49$

119. $(b + 10)^2$

120. $(x + 11)(x - 11)$

121. $\dfrac{x - y}{x}$

122. $\dfrac{r - 2}{r - 3}$

123. $\dfrac{b(b + 3)}{(b - 2)(b - 1)}$

124. 4

125. $\dfrac{3ax}{12x^3}$ and $\dfrac{4a}{12x^3}$

126. $\dfrac{2(b + 1)}{(b + 1)(b - 1)(b + 2)}$ and $\dfrac{3(b + 2)}{(b + 1)(b - 1)(b + 2)}$

127. $\dfrac{7x - 4}{(x + 3)(x - 2)}$

128. $\dfrac{-2(3c + 5)}{(c + 1)(c + 2)(c + 3)}$

129. $y = 2$

130. $y = -\dfrac{13}{3}$

131.

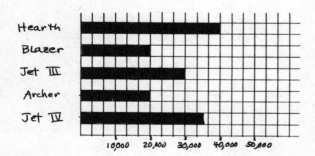

132.
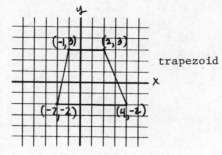
trapezoid

133. IV, II, III, I

134.
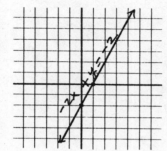

135.
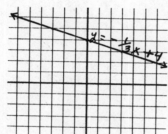

136.
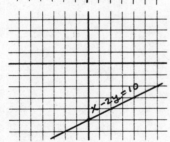

137. Independent, (-1, 1)

138. Inconsistent

139. Independent, (2, -8)

140. Independent, (-5, -2)

141. (-2, 1)

142. -41

143. $(-\frac{3}{5}, -\frac{2}{5})$

144. $(-8, 7)$

145. -10

146. $(2, -2, 6)$

147. y^6

148. $p^4 q^3$

149. $\frac{27x^4 y^8}{4}$

150. 1

151. s^6

152. $\frac{1}{10^2}$

153. x^{10}

154. $p^{1/3} p^{1/3}$

155. $\frac{A^{1/2}}{B^{1/2}}$

156. 9×10^{-5}

157. 0.000023

158. 4.2×10^{-4}, two

159. $50,000$

160. 6

161. 4

162. B^4

163. $4x^2$

164. $x^{3/2}$

165. $\sqrt[3]{B^2}$

166. $\sqrt{m}$

167. $\frac{1}{4}$

168. $3\sqrt{5}$

169. $\frac{\sqrt[3]{4}}{3}$

170. $16y \sqrt[3]{14y}$

171. $\sqrt{65}$

172. $6\sqrt{2} - 6\sqrt{15}$

173. $\sqrt{6}$

174. $\dfrac{5\sqrt{p^3q^3}}{q}$

175. $13\sqrt{6}$

176. $7 \sqrt[3]{9x}$

177. $- \sqrt[3]{x^2y}$

178. $x - \sqrt{x} - 12$

179. $12 \sqrt[3]{4} - 20 \sqrt[3]{2} + 3$

180. $3(\sqrt{2} + 1)$

181. $\dfrac{A + 2\sqrt{AB} + B}{A - B}$

182. $R = \pm 5$

183. No real solution

184. $2\sqrt{35}$

185. $x = 5$ or $x = -1$

186. $R = \dfrac{1}{2}$ or $R = \dfrac{1}{3}$

187. $i = \dfrac{5 \pm \sqrt{37}}{2}$

188. $x = \dfrac{-7 \pm \sqrt{97}}{6}$

189. $12j$

190. $17 + 3j$

191. 20

192. $18 - 48j$

193. $61 - 4j$

194. $\dfrac{6}{5} + \dfrac{3}{5}j$

195. $x = \pm 5j$

196. $i = \dfrac{9}{2} \pm j\dfrac{\sqrt{47}}{2}$

197. $y = -\dfrac{3}{14} \pm \dfrac{j\sqrt{47}}{14}$

198.

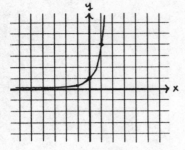

199.

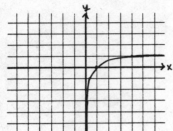

200. $\log_6 216 = 3$

201. $\log_8 1 = 0$

202. $5^1 = 5$

203. $x = 2$

204. $y = -4$

205. 0.8129

206. $0.8639 - 2$

207. 7.43

208. 3.2801

209. 12

210. $x = 9$

211. 0.1635

212. 4.53

213. 2.02

214. IV, $684°$, $-36°$

215. $112.5°$

216. $\sin A = \dfrac{1}{\sqrt{2}}$ $\csc A = \sqrt{2}$
$\qquad\qquad\qquad\qquad \sec A = \sqrt{2}$
$\tan A = 1$ $\cot A = 1$

217. $\cot B = \dfrac{\sqrt{3}}{3}$

218. 0.6330

219. $66°40'$ or $66.67°$

220. $B = 48°$, $a \doteq 6$, $b \doteq 7$

221. $-\dfrac{\sqrt{3}}{2}$, $-\dfrac{1}{2}$, 3, $-\dfrac{2\sqrt{3}}{3}$, -2, $\dfrac{\sqrt{3}}{3}$

222. $(0.35, 3.9)$

223. $(68, 47°)$

224. $9 + 6j$

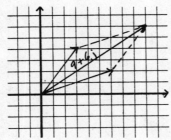

225. $6(\cos 88° + j \sin 88°)$

226. $A = 30°$, $b = 15$, $c = 18$

227. $C = 117°$, $A = 26°$, $B = 36°$

228.

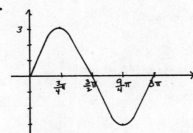

229.

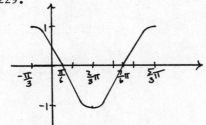

230. $x = 30°$ or $150°$

1.1   1. $804 + 2573 + 71 + 7289 =$ 1._____

2. Show that $(46 + 74) + 31 = 46 + (74 + 31)$ illustrates the associative property of addition of whole numbers.   2._____

3. $7000 - 1653 =$ 3._____

4. $75 + (52 - 19) =$ 4._____

1.2   5. $\begin{array}{r} 2504 \\ \times\ 205 \end{array}$ 5._____

6. $24 \text{ cm} \times 20 \text{ cm} =$ 6._____

7. $738 \div 248 =$ 7._____

8. $72 \div 9 + (49 - 34) =$ 8._____

1.3   9. Write $6 \times 6 \times 6 \times 6$ using exponents.   9._____

10. Write 9^1 without exponents.   10._____

11. Write 10^4 without exponents.   11._____

1.4   12. Find the prime factorization of 575.   12._____

1.5   13. Express $\frac{109}{28}$ as a mixed number.   13._____

14. Express $125\frac{3}{4}$ as an improper fraction.   14._____

1.6   15. Express $\frac{27}{35}$ as an equivalent fraction with a denominator of 105.   15._____

16. Write $\frac{223}{125}$ as an equivalent fraction with a denominator of 375.   16._____

1.7   17. Reduce $\frac{90}{945}$ to lowest terms using prime factorization.   17._____

1.8   18. Write the given fractions as equivalent fractions with the lowest common denominator: 18._____

$$\frac{3}{10}\ ,\ \frac{7}{15}\ ,\ \frac{9}{20}$$

1.9   19. $\frac{7}{8} + \frac{3}{4} + \frac{1}{6} =$ 19._____

20. $3\frac{5}{8} + 4\frac{2}{5}$ 20._____

21. $\frac{7}{10} - \frac{3}{10}$ 21._____

1.10   22. $1\frac{3}{5} \times 4\frac{4}{9}$ 22._____

23. $\frac{6}{7} \div 6$ 23._____

1.11   24. Write 2.504 in words.   24._____

        25. Give the number of significant digits of 0.0020.   25._____

        26. Round 7.4 to units.   26._____

1.12   27. 0.35 + 0.71 =   27._____

        28. 13.5 − 1.673 =   28._____

1.13   Perform the indicated operations. Assume the given data are
        approximations.

        29. 3.1 x 1000   29._____

        30. 1.05 ÷ 0.35   30._____

        31. 11.42 x (0.64 ÷ 0.8) + 0.56.   31._____

1.14   Use a calculator to perform the indicated operations.

        32. $\dfrac{(10 + 7) \times 26}{3}$ 32._____

        33. $\sqrt{176}$ 33._____

1.15   34. Express 4.63 as a fraction.   34._____

        35. Express 0.078 as a fraction.   35._____

        36. Express $\dfrac{3}{8}$ as a decimal.   36._____

1.16   37. Express 2.6 as a percent.   37._____

        38. Express $\dfrac{1}{8}$ as a percent.   38._____

        39. Express 0.90% as a decimal.   39._____

        40. Express $2\dfrac{3}{4}$ % as a decimal.   40._____

        41. Express 225% as a fraction.   41._____

1.17   42. Change 2.5 meters to centimeters.   42._____

        43. Change 400 square decimeters to square meters.   43._____

        44. Change 125 cubic centimeters to cubic decimeters.   44._____

        45. Change 5 quarts to liters.   45._____

2.1   Consider the figure below, given 1 // n, to answer questions
        46 and 47.

46. Name a line intersecting $\overleftrightarrow{DF}$ at point E.

46._____

47. Name three line segments in line m.

47._____

2.2 48. Find the complement of an angle whose measure is 14°.

48._____

49. Find the supplement of an angle whose measure is 112°.

49._____

50. Give the degree measure of $\angle\beta$ in the following figure:

50._____

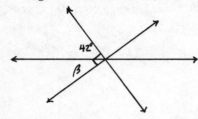

2.3 51. Find the perimeter of the following triangle:

51._____

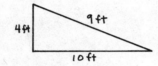

2.4 52. Find the perimeter and area of the following polygon:

52._____

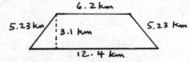

2.5 53. Find the circumference and area of a circle whose radius is 5 cm.

53._____

2.6 54. Find the surface area of a cube if one of its edges is 5 meters.

54._____

55. Find the volume of a sphere with diameter 7.4 meters.

55._____

3.1 56. Use a signed number to represent the drop in voltage of sixteen.

56._____

57. Insert the proper sign, $>$ or $<$ between the numbers:

57._____

$$-7 \qquad -9$$

58. Evaluate $\left|-23\right|$.

58._____

3.2 59. Add the following real numbers. Check the sum by adding on the number line.

59._____

$$(-4) + 5$$

60. Add: $(-20) + (-2)$

60._____

61. Add: $(7) + (-2) + (-8) + 13$

61._____

3.3   62. Subtract: $10 - 6$ 62._____

         63. Subtract: $-4 - 7$ 63._____

         64. Perform the indicated operations: 64._____

$$10 - 8 - 20 - (-1) - (-5)$$

3.4   Add or subtract the following terms:

         65. $14A + 9A$ 65._____

         66. $-5d - (-7d)$ 66._____

         67. $-18V - 20V$ 67._____

     Add or subtract the following polynomials:

         68. $(10Y - 11) - (-2Y - 5)$ 68._____

         69. $(2R^2 + 3R - 9) + (3R^2 - 2R + 9)$ 69._____

3.5   Perform the indicated operations:

         70. $(5)(-5)$ 70._____

         71. $(-55) \div (-5)$ 71._____

         72. $-48 \div \sqrt{64} - 81 - 3^2$ 72._____

3.6   73. $m^3 \cdot m$ 73._____

         74. $(4B)(-2B^2)(10B^4)$ 74._____

         75. $-2x^2y(4xy - 3x^2y^2)$ 75._____

3.7   76. $x^9 \div x$ 76._____

         77. $\dfrac{d}{d^4}$ 77._____

3.8   78. $2R - \{4 - (9R - 1)\}$ 78._____

         79. $6e - \{3e - 5e(3e - 9)\}$ 79._____

         80. $-3Z[4Z - 6(Z + 1) + 3Z(Z - 2)] + 2(4 - Z^2)$ 80._____

3.9   81. Evaluate $9c - 8$ if $c = -3$.   81._____

         82. If $x = 5$ and $y = -4$, find the value of $\dfrac{7(-2x + 6y)}{20x}$ 82._____

         83. How many feet of fence is needed to put a fence around   83._____
            a rectangular yard that measures 120 feet by 100 feet?

4.1   Solve each equation for the given variable:

         84. $a - 6 = 13$ 84._____

         85. $-\dfrac{2}{5}x = -10$ 85._____

86. $4 - c = 10$ 86._____

4.2   87. $5y + 17 = y - 7$ 87._____

88. $\frac{x}{10} + \frac{2}{5} = \frac{17}{5}$ 88._____

89. $4(b + 6) - 18 = 2(b + 4)$ 89._____

4.3   Solve the following inequalities and graph the solution
       on the number line:

90. $X - 4 > -9$ 90._____

91. $6t - 5 \leqslant 5t - 5$ 91._____

92. $2(3e - 1) \geqslant 4(3e - 5)$ 92._____

4.4   93. Solve $E = IR$ for I.   93._____

94. Solve $V = V_0 + at$ for t.   94._____

95. Solve $2V_a - V_1 = V_2$ for V_a, and find V_a given $V_1 = 66.04$ 95._____

ft/sec, and $V_2 = 77.5$ ft/sec.

4.5   Translate each phrase into an algebraic expression. Let x
       represent the unknown number.

96. The sum of 15 and a number.   96._____

97. One half of a number subtracted from the number.   97._____

98. Translate into an algebraic equation with x as the   98._____
     unknown number, and solve to find the unknown number:

     the difference of a number and −8 is 24.

4.6   Solve and check each of the following problems:

99. One resistor in a circuit is 1.8 times another. If   99._____
     3.2 ohms is subtracted from the larger resistor, it
     is equal to the smaller. Find the values of the
     resistors.

100. The larger base of a trapezoid is 6 inches longer than   100._____
      the smaller base. If the two sides total 10 inches
      and the perimeter of the trapezoid is 20 inches, find
      the length of the two bases.

101. How many liters of a 30% solution of acid should be   101._____
      added to 50 liters of a 90% solution of acid to obtain
      a 60% solution?

4.7   102. Express each of the following as a ratio in lowest terms: 102._____

      a. 18 cm to 38 cm   b. 22%

103. A cement block weighs 80 kg and a steel block weighs 1640   103._____
      kg. What is the ratio of the steel block to the cement
      block?

104. Find the resistance of 250 m of copper wire if the resistance of 30 m of the same copper wire is 4.2 ohms.

104._____

4.8   105. Write "D varies directly as M" as an equation containing a constant of proportionality, k.

105._____

106. If y varies inversely as x, and y = 18 when x = 24, find the constant of variation.

106._____

107. The currents in a parallel circuit are inversely proportional to the resistances. If the current I_1 is 0.6A when the resistance R_1 is 140 ohm,

find the current, I_2, when the resistance R_2 is 120 ohms.

107._____

5.1   Factor:

108. $3R^3 + 19R^9$

108._____

109. $12a^8b^9 - 36a^7b^8 - 44a^6b^7$

109._____

5.2   Multiply:

110. $(m + 5)(m - 11)$

110._____

111. $(3a + 8)(3a - 8)$

111._____

5.3   Factor:

112. $x^2 - 5x + 6$

112._____

113. $j^2 + 11j + 24$

113._____

114. $2C^2 - 40C + 182$

114._____

5.4   Factor:

115. $6Y^2 - 5Y - 1$

115._____

116. $8X^2 + 30XY + 25Y^2$

116._____

117. $12D^2 - 21D + 9$

117._____

5.5   118. Square the binomial, $(2j - 9k)^2$

118._____

119. Factor: $9i^2 - 24i + 16$.

119._____

120. $144m^2 - 1$

120._____

6.1   Reduce to lowest terms:

121. $\dfrac{4x - 8}{x^2 - 4x + 4}$

121._____

122. $\dfrac{x^2 - 2xy + y^2}{x^2 - y^2}$

122._____

123. $\dfrac{3x^2 + 6x + 3}{15x^2 - 15}$

123._____

6.2   Perform the indicated operations:

124. $\dfrac{2a^2 - 13a - 7}{a^2 - 6a - 7} \cdot \dfrac{a^2 - a - 2}{2a^2 - 5a - 3}$

124._____

125. $\dfrac{4a^2 + 12a + 9}{3a^2 + 2a - 1} \div \dfrac{4a^2 - 9}{3a^2 + 14a - 5}$

125._____

6.3   126.   Find the L.C.D. of the following algebraic
              fractions, and convert to equivalent fractions
              with a common denominator.

126._____

$$\dfrac{5}{3y + 15}, \quad \dfrac{y}{y^2 - 25}$$

6.4   Perform the indicated operations and simplify the
       answers:

127. $\dfrac{x - 4}{8x} - \dfrac{x - 2}{4x^2}$

127._____

128. $\dfrac{5}{3y + 15} + \dfrac{y}{y^2 - 25}$

128._____

6.5   Solve and check:

129. $\dfrac{2}{x + 2} = \dfrac{5}{8}$

129._____

130. $\dfrac{3}{y^2 - 1} - \dfrac{1}{y - 1} = \dfrac{1}{y + 1}$

130._____

7.1   131.   To test the maximum output of the different stoves,
              80 stoves were tested: 16 Hearth, 17 Blazer, 14
              Jet IV, 17 Arnold, and 16 Jet III. Compute the
              percentage that each part is of the whole.

131._____

132.   Draw a circle graph using the information from
       problem 131.

132._____

7.2   133.   Given the relation $\{(4, -2), (-1, 3), (-2, -2),$
              $(2, 3)\}$, state the domain and range of the relation.

133._____

134.   Is the relation in problem 133 a function?

134._____

7.3   135.   Graph the equation $4x + y = 3$.

135._____

196

7.4   136. Use the slope-intercept method to graph the   136._____
                  following linear equation:

                        $y = \frac{3}{4} x + 1$

8.1   Solve the following systems by graphing, and tell whether
             they are independent, inconsistent, or dependent. If
             independent, give the solution.

             137. $x - 3y = -9$ 137._____

                   $2y = \frac{2}{3} x + 6$

             138. $4X + Y = -10$ 138._____
                   $3X + 4Y = -1$

             139. $5d - 3e = -11$ 139._____
                   $7d + 3e = -1$

8.2   140. Solve the following system by using the addition   140._____
                   method. Tell whether the system is independent,
                   inconsistent, or dependent. If independent, give
                   the solution.

                   $3a + b = 2$
                   $5a + 4b = 15$

             141. Solve the following system by substitution: 141._____

                   $M - 3N = 7$
                   $-2M + 6N = -14$

8.3   142. Evaluate: $\begin{vmatrix} 8 & 6 \\ 4 & 3 \end{vmatrix}$ 142._____

             Solve the following systems by using Cramer's Rule:

             143. $x - y = 2$ 143._____
                   $3x - 6y = 10$

             144. $3R + 3T = 3$ 144._____
                   $5R + 3T = 4$

8.4   145. Evaluate: $\begin{vmatrix} -3 & 5 & 1 \\ 3 & -5 & 1 \\ -3 & -5 & -1 \end{vmatrix}$ 145._____

             146. Solve by using Cramer's Rule: 146._____

                   $3x + 3y + z = 5$
                   $3x + 2y - 4z = -22$
                   $2x + 5y - z = 1$

9.1   Simplify:

147. $2^4 \cdot 2^2 \cdot 2$ 147._____

148. $(R^4 S^2)(R^3 S^5)$ 148._____

149. $(x^2 y^3)^2 (xy^4)^3$ 149._____

9.2   Simplify:

150. $(-4y)^0$ 150._____

151. $\dfrac{(-2pq)^0}{(pq)^3}$ 151._____

9.3   Simplify and give answers without negative exponents:

152. $4s^{-1}$ 152._____

153. $\left(\dfrac{x^3}{y^4}\right)^{-2}$ 153._____

9.4   Simplify and express answers with positive exponents only:

154. $\dfrac{r^{6/7}}{r^{3/7}}$ 154._____

155. $\dfrac{x^{-3/5} y^{-1/7}}{x^{1/5} y^{5/7}}$ 155._____

9.5   156. Express 6,500,000 in scientific notation.   156._____

157. Express 8×10^6 as an ordinary number.   157._____

158. Express 500,000 in scientific notation and indicate the   158._____
      number of significant digits.

9.6   159. Use scientific notation to calculate: 159._____

$$\frac{0.00025}{500,000}$$

10.1   Find the following roots:

160. $\sqrt[10]{1}$ 160._____

161. $-\sqrt[3]{-27}$ 161._____

10.2   Assume that all variables represent nonnegative numbers, and
           find the roots:

162. $\sqrt[3]{B^9}$ 162._____

163. $\sqrt{a^4 b^2 c^8}$ 163._____

164. Express $\sqrt[5]{y^2}$ in exponential form.   164._____

165. Express $4^{1/4}$ in radical form.

165._____

166. Reduce the index of $\sqrt{R^2}$.

166._____

167. Evaluate $\left(\dfrac{-8}{27}\right)^{-1/3}$

167._____

10.3   Simplify the following:

168. $\sqrt{75}$

168._____

169. $\sqrt{32x^3}$

169._____

170. $-\sqrt{27a^5b^4}$

170._____

10.4   Multiply the following radicals. Simplify wherever possible.

171. $-\sqrt[3]{7} \cdot \sqrt[3]{6}$

171._____

172. $\sqrt[3]{x^3y} \cdot \sqrt[3]{8x^2y^2} \cdot \sqrt[3]{3x^4y}$

172._____

10.5   Simplify:

173. $\dfrac{\sqrt[3]{24a^4b^4}}{\sqrt[3]{3ab}}$

173._____

174. $\dfrac{6}{\sqrt{3}}$

174._____

10.6   Add or subtract the following radicals:

175. $8\sqrt{10} - 3\sqrt{10}$

175._____

176. $4\sqrt{3} + 2\sqrt{5} - 2\sqrt{3} - 5$

176._____

177. $3\sqrt{72r^2} + 2\sqrt{32r^2} - 3\sqrt{18r^2}$

177._____

10.7   Multiply and simplify answers when possible:

178. $(2 - \sqrt{R})(5 + 2\sqrt{R})$

178._____

179. $(\sqrt{3} - 4)^2$

179._____

10.8   Rationalize:

180. $\dfrac{5}{3 - \sqrt{3}}$

180._____

181. $\dfrac{\sqrt{3} + 1}{\sqrt{2} - 2}$

181._____

11.1   Solve by taking the square root of both members of the equation.

182. $a^2 = 45$

182._____

183. $(b + 5)^2 = 12$

183._____

184. Find the distance between the points (2, 8) and (-1, 3).

184._____

11.2   Solve the following by completing the square:

185. $y^2 - 3y - 7 = 0$

185._____

186. $2e^2 - e - 5 = 0$

186._____

11.3   Solve the following equations using the quadratic formula:

187. $j^2 + 11j + 3 = 0$

187._____

188. $2R^2 - R = 11$

188._____

11.4   189. Simplify: $-2\sqrt{-36}$

189._____

Add or subtract the given complex numbers. Express answers in the form a + bj.

190. $(-2 + j) - (6 - 3j)$

190._____

191. $(5 - 3j\sqrt{2}) + (8 - 2j\sqrt{2})$

191._____

11.5   Perform the indicated operations and express answers in the form of a + bj.

192. $j(8 - j)$

192._____

193. $(4 + j)(2 + 3j)$

193._____

194. $\dfrac{4 + 3j}{7 - 5j}$

194._____

11.7   Solve for the given variable and express answers in the form of a + bj.

195. $w^2 = -147$

195._____

196. $9c^2 + 2c = -1$

196._____

197. $8x^2 = -x - 1$

197._____

12.1   198. Graph the following exponential function: $y = \left(\dfrac{1}{4}\right)^x$

198._____

199. Graph the following logarithmic function: $x = \left(\dfrac{1}{4}\right)^y$

199._____

12.2   200. Express $10^5 = 100{,}000$ in logarithmic form.

200._____

201. Express $(5)^{-2} = \dfrac{1}{25}$ in logarithmic form.

201._____

202. Express $\log_{11} 121 = 2$ in exponential form.

202._____

203. Solve $\log_4 \dfrac{1}{64} = x$ for x.

203._____

204. Solve $\log_4 y = -2$ for y.

204._____

12.3   Find the logarithm of each of the following:

       205.   log 38,900   205._____

       206.   log 0.4   206._____

12.4   207.   Find the antilog of 3.2148.   207._____

12.5   208.   Use the properties of logarithms to find   208._____

           $\log \dfrac{0.0065}{0.0432}$

       209.   Solve $\log (x + 1) = \log x + \log 3$ for x.   209._____

       210.   Solve $\log (5x + 2) = \log (x + 5) - \log 2$ for x.   210._____

12.6   211.   Compute $\dfrac{0.135}{24.3}$ using logarithms.   211._____

12.7   Find the natural logarithms of the following:

       212.   ln 425   212._____

       213.   ln 0.75   213._____

13.1   214.   State the quadrant of 243°. Give one positive   214._____
           and one negative coterminal angle for 243°.

       215.   Convert 136° to radian measure.   215._____

13.2   216.   Given: right triangle ACB with C = 90°, and   216._____
           cot B = $\dfrac{7}{5}$, give the other five trigonometric

           ratios of angle B.

       217.   Given: right triangle ACB, C = 90°, and   217._____

           sin A = $\dfrac{3}{\sqrt{10}}$, verify $\sin^2 A + \cos^2 A = 1$.

13.3   218.   Use Table 3 or a calculator to find cos 84°30'.   218._____

       219.   Use Table 3 or a calculator to find x if   219._____
           $0° \leqslant x < 90°$, and tan x = 0.2192.

13.4   220.   The angle of elevation to the top of a tree is   220._____
           29°30' from a point 20.5 feet from the base of
           the tree. Find the height of the tree.

13.5   221.   Find θ where $0° \leqslant \theta < 360°$, given sin θ = -0.4383,   221._____
           and tan $\theta > 0$.

13.6   222.   Change the vector (7.02, 113°) from polar to   222._____
           rectangular form.

       223.   Change the vector (8.1, 5.6) from rectangular to   223._____
           polar form.

13.7   224. Multiply the vector $5 + j$ by j and graph both   224._____
              vectors in the complex plane.

        225. Divide $12(\cos 140° + j \sin 140°)$ by   225._____
              $4(\cos 70° + j \sin 70°)$.

13.8   226. Use the Law of Sines to solve the triangle, given: 226._____

              $A = 88°$
              $a = 13$
              $b = 18$

13.9   227. Solve the oblique triangle, given: 227._____

              $a = 50$
              $b = 74$
              $C = 52°$

13.10   228. Graph one period of $y = 2 \cos 4x$ beginning at $x = 0$. 228._____

13.11   229. Graph one period of $y = 2 \sin (3x - 2\pi)$.   229._____

13.12   230. Solve the equation $3 \cos^2 x + \cos x - 1 = 0$ for   230._____
              $0° \leqslant x < 360°$.

FINAL EXAMINATION ANSWERS
FORM B

1. 10,737

2. (46 + 74) + 31 = 46 + (74 + 31)
   120   + 31 = 46 +   105
              151 = 151

3. 5347

4. 108

5. 513,320

6. 480 sq cm

7. 2 R 242

8. 23

9. 6^4

10. 9

11. 10,000

12. 5^2 x 23

13. $3 \frac{25}{28}$

14. $\frac{503}{4}$

15. $\frac{81}{105}$

16. $\frac{669}{375}$

17. $\frac{2}{21}$

18. $\frac{18}{60}$, $\frac{28}{60}$, $\frac{27}{60}$

19. $1 \frac{19}{24}$

20. $8 \frac{1}{40}$

21. $\frac{4}{10}$ or $\frac{2}{5}$

22. $7 \frac{1}{9}$

23. $\frac{1}{7}$

24. two and five hundred four thousandths

25. two

26. 7

27. 1.06

28. 11.8

29. 3000

30. 3.0

31. 10

32. 147.33

33. 13.266

34. $\frac{463}{100}$

35. $\frac{39}{500}$

36. 0.375

37. 260%

38. 12.5%

39. 0.009

40. 0.0275

41. $\frac{9}{4}$

42. 250 cm

43. 4 m^2

44. 0.015 dm^3

45. 4.731

46. line m

47. $\overline{BE}$, $\overline{BG}$, and $\overline{EG}$

48. 76°

49. 68°

50. 48°

51. 23 ft

52. P = 29.1 km, A ≐ 29 km^2

53. C = 31.4 cm, A = 78.5 cm^2

54. 150 m^2

55. V ≐ 210 m^3

56. -16

57. $>$

58. 23

59.

       $(-4) + 5 = 1$

60. -22

61. 10

62. 4

63. -11

64. -12

65. $23A$

66. $2D$

67. $-38V$

68. $12Y - 6$

69. $5R^2 + R$

70. -25

71. 11

72. -96

73. M^4

74. $-80B^7$

75. $-8X^3Y^2 + 6X^4Y^3$

76. X^8

77. $\dfrac{1}{d^3}$

78. $11R - 5$

79. $21e^2 - 48e$

80. $-9Z^3 + 22Z^2 + 18Z + 8$

81. -35

82. $-\dfrac{119}{50}$

83. 440 feet

84. $a = 19$

85. $x = 25$

86. $c = -6$

87. $Y = -6$

88. $X = 30$

89. $b = 1$

90. $x > -5$

91. $t \leqslant 0$

92. $e \leqslant 3$

93. $I = \dfrac{E}{R}$

94. $t = \dfrac{V - V_0}{a}$

95. $V_a = 71.77$ ft/sec

96. $15 + X$

97. $X - \dfrac{1}{2} X$

98. $X - (-8) = 24$, $X = 16$

99. 4 ohms and 7.2 ohms

100. 2 in., 8 in.

101. 50 liters

102. a. $\dfrac{9}{19}$ b. $\dfrac{11}{50}$

103. 41 to 2

104. 35 ohms

105. $D = km$

106. $Y = 432$

107. 0.7 A

108. $R^3(3 + 19R^6)$

109. $4a^6 b^7(3a^2 b^2 - 9ab - 11)$

110. $6H^2 + 37H + 45$

111. $9a^2 - 64$

112. $(x - 2)(x - 3)$

113. $(j + 3)(j + 8)$

114. $2(c - 13)(c - 7)$

115. $(6y + 1)(y - 1)$

116. $(4x + 5y)(2x + 5y)$

117. $3(4D - 3)(D - 1)$

118. $4j^2 - 36jk + 81k^2$

119. $(3i - 4)^2$

120. $(12m + 1)(12m - 1)$

121. $\dfrac{4}{x - 2}$

122. $\dfrac{x - y}{x + y}$

123. $\dfrac{3(x + 1)}{5(x - 1)}$

124. $\dfrac{a - 2}{a - 3}$

125. $\dfrac{(2a + 3)(a + 5)}{(a + 1)(2a - 3)}$

126. $\dfrac{5(y - 5)}{3(y + 5)(y - 5)}$ and $\dfrac{3y}{3(y + 5)(y - 5)}$

127. $\dfrac{x^2 - 6x + 4}{8x^2}$

128. $\dfrac{8y - 25}{3(y^2 - 25)}$

129. $x = \dfrac{6}{5}$

130. $y = \dfrac{3}{2}$

131.

Hearth	20%
Blazer	21%
Jet IV	18%
Arnold	21%
Jet III	20%

132.

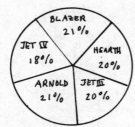

133. $D = \{-2, -1, 2, 4\}$, $\{R = -2, 3\}$

134. Yes

135.

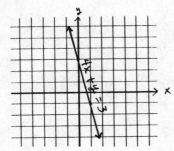

136.

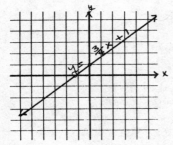

137. Dependent

138. Independent, $(-3, 2)$

139. Independent, $(-1, 2)$

140. Independent, $(-1, 5)$

141. Dependent

142. 0

143. $\left(\dfrac{2}{3}, -\dfrac{4}{3}\right)$

144. $\left(\dfrac{1}{2}, \dfrac{1}{2}\right)$

145. -60

146. $(-2, 2, 5)$

147. 2^7

148. $R^7 S^7$

149. $x^7 y^{18}$

150. 1

151. $\dfrac{1}{p^3 q^3}$

152. $\dfrac{4}{S}$

153. $\dfrac{y^8}{x^6}$

154. $r^{3/7}$

155. $\dfrac{1}{x^{4/5} y^{6/7}}$

156. 6.5×10^6

157. $8,000,000$

158. 5×10^5, 1

159. 5×10^{-10}

160. 1

161. 3

162. B^3

163. $a^2 b c^4$

164. $y^{2/5}$

165. $\sqrt[4]{4}$

166. R

167. $-\dfrac{3}{2}$

168. $5\sqrt{3}$

169. $4x\sqrt{2x}$

170. $-3a^2 b^2 \sqrt{3a}$

171. $-\sqrt[3]{42}$

172. $2x^3 y^3 \sqrt{3y}$

173. $2ab$

174. $2\sqrt{3}$

175. $5\sqrt{10}$

176. $2\sqrt{3} + \sqrt{5}$

177. $17r \sqrt{2}$

178. $10 - \sqrt{R} - 2R$

179. $19 - 8\sqrt{3}$

180. $\dfrac{5(3 + \sqrt{3})}{6}$

181. $\dfrac{\sqrt{6} + \sqrt{2} + 2\sqrt{3} + 2}{-2}$

182. $a = \pm 3\sqrt{5}$

183. $b = -5 \pm 2\sqrt{3}$

184. $\sqrt{34}$

185. $y = \dfrac{3 \pm \sqrt{37}}{2}$

186. $3 = \dfrac{1 \pm \sqrt{41}}{4}$

187. $j = \dfrac{-11 \pm \sqrt{109}}{2}$

188. $R = \dfrac{1 \pm \sqrt{89}}{4}$

189. $-12j$

190. $-8 + 4j$

191. $-5j + 2$

192. $1 + 8j$

193. $5 + 14j$

194. $\dfrac{13}{74} + \dfrac{41}{74} j$

195. $W = \pm 7j \sqrt{3}$

196. $c = -\dfrac{1}{9} \pm \dfrac{j\sqrt{31}}{9}$

197. $x = -\dfrac{1}{16} \pm \dfrac{j\sqrt{31}}{16}$

198.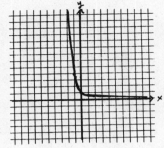

199.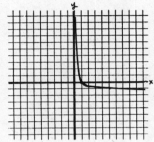

200. $\log_{10} 100,000 = 5$

201. $\log_5 \frac{1}{25} = -2$

202. $11^2 = 121$

203. $x = -3$

204. $y = \frac{1}{16}$

205. 2.5899

206. 0.6021

207. 1640

208. 0.1774−1

209. $x = \frac{1}{2}$

210. $x = \frac{1}{9}$

211. 0.0056

212. 6.05

213. −0.29

214. III, $603°$, $-117°$

215. $\frac{34\pi}{45}$

216. $\sin B = \dfrac{5\sqrt{74}}{74}$ $\qquad \csc B = \dfrac{\sqrt{74}}{5}$

$\cos B = \dfrac{7\sqrt{74}}{74}$ $\qquad \sec B = \dfrac{\sqrt{74}}{7}$

$\tan B = \dfrac{5}{7}$

217. $\left(\dfrac{3}{\sqrt{10}}\right)^2 + \left(\dfrac{1}{\sqrt{10}}\right)^2 = \dfrac{9}{10} + \dfrac{1}{10} = \dfrac{10}{10} = 1$

218. 0.0958

219. $12°20'$ or $12.36°$

220. 11.6 ft

221. $206°$

222. $(-2.74, 6.46)$

223. $(9.8, 35°)$

224. $-1 + 5j$

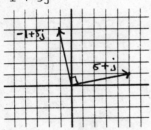

225. $3(\cos 70° + j \sin 70°)$

226. No solution

227. No solution

228.

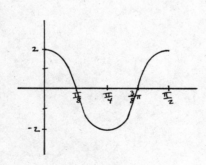

229.

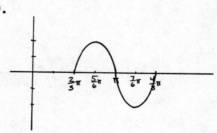

230. $x = 64°, 296°, 140°,$ or $220°$

Section 12.1

2.
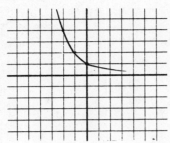

4.
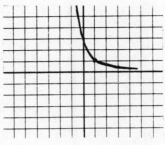

6.
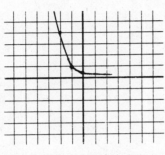

8.
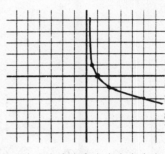

10.
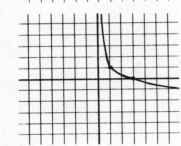

12.
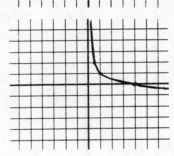

14. 520

Section 12.2

2. $\log_3 1 = 0$ 4. $\log_{10} \frac{1}{100} = -2$ 6. $\log_3 9 = 2$ 8. $\log_2 16 = 4$ 10. $2^5 = 32$

12. $10^3 = 1,000$ 14. $3^{-2} = \frac{1}{9}$ 16. $\left(\frac{1}{2}\right)^0 = 1$ 18. 2   20. $\frac{1}{2}$ 22. 4

24. 1   26. 0   28. −2   30. 2   32. −2   34. $\frac{1}{100}$

Section 12.3

2. 0.7832   4. 0.0414   6. 0.3010   8. 2.7067   10. 0.7582−1   12. 4.6721

14. 0.7084−2   16. 1.4624   18. −12   20. 0.6785−10   22. 3.7419

Section 12.4

2. 1.27   **4.** 6.46   **6.** 5.07   **8.** 6020   **10.** 0.00735   **12.** 5,140,000

14. 0.000703   **16.** 8,110,000,000   **18.** 300,000,000 m/s   **20.** 0.00000002 cm

Section 12.5

2. 0.3765-5   **4.** 0.4835-1   **6.** 0.8225-2   **8.** 0.8644-6   **10.** 0.4337-15

12. 52.4583   **14.** 2.4328   **16.** 0.2897-1   **18.** 0.9958-1   **20.** 0.0168

22. 0.2888-1   **24.** 0.2090   **26.** 2.446   **28.** 0.8108   **30.** $x = 6$ **32.** $x = 1$

34. $x = 5$ **36.** $2\log V - \log g$ **38.** $(\log P + \log T) - \log V$

40. $\log 20 + (\log E_2 - \log E_1)$

Section 12.6

2. 5.72×10^7 **4.** 4.26×10^{-2} **6.** 1.36×10^{-2} **8.** 1.61   **10.** 1.76×10^{14}

12. 1.80×10^{-6} **14.** 1.41   **16.** 9.22   **18.** 3.38×10^{-1} **20.** 4.42

22. 6.42×10^2 **24.** 3.2×10^5 **26.** 1.06×10^6

Section 12.7

2. 4.39   **4.** 6.64   **6.** 1.47   **8.** 4.52   **10.** -0.198   **12.** 46.05